Introduction to Minerals and Rocks

Joseph C. Cepeda

West Texas A&M University

Macmillan College Publishing Company
New York

Maxwell Macmillan Canada
Toronto

Maxwell Macmillan International
New York   Oxford   Singapore   Sydney

Cover photo: Grant Heilman Photography, Inc. Image is of Mono Lake, California.
Editor: Robert A. McConnin
Production Editor: Julie Anderson Tober
Photo Editor: Chris Migdol
Cover Designer: Robert Vega
Production Manager: Patricia A. Tonneman
Electronic Text Management: Marilyn Wilson Phelps, Matthew Williams, Jane Lopez, Vincent A. Smith
Illustrations: Precision Graphics

This book was set in Latin by Macmillan College Publishing Company and was printed and bound by R. R. Donnelley & Sons Company. The cover was printed by Phoenix Color Corp.

Copyright © 1994 by Macmillan College Publishing Company, Inc.
Printed in the United States of America

All rights reserved. No part of this book may be reproduced or transmitted in any form or by any means, electronic or mechanical, including photocopy, recording, or any information storage and retrieval system, without permission in writing from the Publisher.

Macmillan College Publishing Company
866 Third Avenue
New York, New York 10022

Macmillan College Publishing Company is part of the
Maxwell Communication Group of Companies.

Maxwell Macmillan Canada, Inc.
1200 Eglinton Avenue East, Suite 200
Don Mills, Ontario M3C 3N1

Library of Congress Cataloging-in-Publication Data

Cepeda, Joseph C.
    Introduction to minerals and rocks / Joseph C. Cepeda.
      p. cm.
    Includes bibliographical references and index.
    ISBN 0-02-320452-4
    1. Minerals. 2. rocks. I. Title.
QE363.2.C45   1994
552—dc20   93-31893
                                                    CIP

Printing: 1 2 3 4 5 6 7 8 9 Year: 4 5 6 7

Photos by the author unless otherwise indicated.

Preface

Rocks and minerals are the most abundant objects on the surface of Earth. Our curiosity about them is often stymied by a lack of knowledge about their formation and the techniques for studying them. Despite the advances in analytical techniques in the last 40 years, the first step in deciphering the history of a rock or mineral is still the collection and examination of a hand specimen. Only in a hand-specimen are the texture, structure, color, and other characteristics of the rock or mineral evident.

This book tells you what to look for, how to observe, and how to identify the common minerals and rocks. No special tools are necessary, just a hand lens or magnifying glass, and a few others such as a pocket knife, a penny, and a glass plate. A small field guide to minerals would also be useful in providing descriptions that are beyond the scope of this text.

This book is intended as an introductory text for a college sophomore- or junior-level laboratory course in rocks and minerals. No chemistry or mathematics beyond high school chemistry or mathematics is necessary. This text provides a glimpse at the various types of rocks and minerals to encourage the student to learn more about this fascinating subject.

Chapter 1 is a brief historical narrative of the importance of minerals in the evolution of our civilization. Here the student will discover the vital and widespread use of minerals in this age of plastics.

Chapter 2 is an introduction, or review, of elementary chemical principles and a discussion of the physical properties of minerals. These physical properties allow us to nondestructively determine the properties of mineral specimens. The chapter concludes with a classification scheme for minerals and enables the student to see how mineralogists organize and classify the more than 3,000 mineral species.

Chapter 3 describes the ore minerals, and discusses how they form and are concentrated to form economic ore deposits. The chapter concludes with a discussion of mineral associations. Chapter 4 describes the industrial minerals, which are vital to industrial processes from steel-making to fertilizers. Chapter 5 describes the rock-forming minerals that form the major portion of Earth's crust. Chapter 6 is an introduction to the planet Earth. It gives us a brief historical perspective of the geological detective work necessary to determine the composition of Earth's interior. This sets the stage for our study of rocks.

Chapter 7 describes one of the most spectacular of natural phenomena—volcanoes. Volcanic rocks have proven to be incredibly versatile sources of information by providing data on the history of Earth's magnetic field, the age of volcanic eruptions, and the movement of the continents through geologic time.

Chapter 8 is an introduction to the world of plutonic igneous rocks, which crystallized from molten material deep underground and are interpreted on the basis of the clues preserved in the rock. They are also the source of most of the ore and industrial minerals discussed in Chapters 3 and 4.

Chapter 9 begins a discussion of the sedimentary rocks, the most abundant rocks on Earth's surface. The clues written in their layers record the uplift of mountains and the collision of continents. The sedimentary rocks also contain the fossil record which traces the evolution of life on the planet over a time span of more than 3 billion years.

Chapter 10 concludes our story on sedimentary rocks by describing the formation of the carbonate rocks and related evaporites.

Chapter 11 describes the metamorphic rocks, the most challenging of all. These rocks form in a variety of environments, and the geologist must not only look at their mineralogy to determine the conditions of metamorphism, but also look past and beyond the metamorphism to determine the rock type prior to the metamorphism.

I would like to thank the following reviewers of the manuscript for their input: Robert A. Christman, Western Washington University; Robert J. Foster, San Jose State University; Cornelis Klein, University of New Mexico; and Judith B. Moody, J. B. Moody and Associates (Athens, Ohio).

Joseph Cepeda

Contents

1 Minerals in the Past and Present World *1*

    1.1 The Use of Minerals by Early Civilizations *2*
    1.2 The Quest for Ore Deposits *2*
    1.3 Elemental Abundances in the Earth's Crust *4*

2 The Chemical and Physical Properties of Minerals *7*

    2.1 Structure of the Atom *7*
    2.2 Ions and Isotopes *9*
    2.3 Chemical Bonding *10*
    2.4 Introduction to the Realm of Minerals *13*
    2.5 Physical Properties of Minerals *14*
           Color *15*
           Crystal Form and Habit *15*
           Cleavage and Fracture *15*
           Hardness *17*
           Luster *18*
           Specific Gravity *18*
           Streak *18*
           Other Physical Properties *18*
    2.6 Chemical Classification of Minerals *19*

3 The Ore Minerals: Native Elements, Sulfides, and Sulfosalts *21*

    3.1 The Metallic Native Elements *21*
           Silver *22*
           Gold *25*
           Copper *26*
           Platinum *27*

3.2 Nonmetallic Native Elements 28
    Sulfur 28
    Diamond and Graphite 28
3.3 Sulfides and Sulfosalts 29
    Physical Properties 32
3.4 Formation of a Hydrothermal Ore Deposit 33
    The Santa Eulalia District, Chihuahua, Mexico 33
    The Coeur d'Alene District, Idaho 35
    The Cripple Creek District, Colorado 35
3.5 The Concept of Mineral Associations 36

4 The Industrial Minerals 37

4.1 The Oxides and Hydroxides 37
    Corundum 40
    Hematite and Magnetite 41
    Ilmenite and Rutile 41
    Cassiterite 42
    Other Oxide Minerals 43
    The Hydroxide Minerals 43
4.2 The Carbonates 45
4.3 The Sulfates 48
4.4 The Borates 50
4.5 The Halide Minerals 53
4.6 The Tungstates, Molybdates, Phosphates, and Vanadates 54

5 The Rock-Forming Minerals: The Silicates 59

5.1 Crystal Structure 59
5.2 The Nesosilicates 62
5.3 The Sorosilicates 67
5.4 The Cyclosilicates 69
5.5 The Inosilicates 70
    The Pyroxene Group 71
    The Amphibole Group 74
    The Pyroxenoid Group 75
5.6 The Phyllosilicates 76
    The Serpentine Subgroup 76
    The Mica Subgroup 77
    The Clay Mineral Subgroup 79
5.7 The Tectosilicates 80
    The Silica Subgroup 80
    The Feldspar Subgroup 85
    The Feldspathoid Subgroup 87
    The Zeolite Subgroup 87

Contents

6 The Structure and Composition of the Earth *89*
   6.1 Sources of Information *89*
   6.2 Exploring the Earth's Interior—A Historical Perspective *90*
   6.3 Evidence from the Meteorites *93*
   6.4 Plate Tectonics and the Origin of the Crust *95*
         The Mechanism of Plate Tectonics *95*
   6.5 Experimental Studies of Partial Melting *96*

7 Volcanism and Volcanic Rocks *99*
   7.1 Introduction *99*
   7.2 Scientific Implications of Volcanic Rocks *99*
         Plate Tectonics *99*
         Radiometric Dating *100*
         History of Earth's Magnetic Field *100*
         Hotspot Tracers *101*
   7.3 Chemistry and Physics of Magma *102*
   7.4 Types of Volcanoes and Vent Areas *103*
         Shield Volcanoes *103*
         Composite Volcanoes *110*
         Cinder Cones *116*
         Collapse Calderas *117*
   7.5 Ash Flows and Ash-Flow Tuffs *118*
   7.6 Structures and Textures of Lava Flows *121*
   7.7 Classification of Volcanic Rocks *123*

8 Granites and Other Plutonic Rocks *127*
   8.1 Geometry of Plutonic Bodies *127*
   8.2 Composite Granitic Batholiths *128*
         The Sierra Nevada Batholith *129*
   8.3 Pegmatites *131*
         Structure and Mineralogy of Pegmatites *131*
         Pegmatite Formation *133*
   8.4 Stratiform Complexes *133*
   8.5 Alpine-Type Ultramafic Bodies *135*
   8.6 Anorthosites *135*
   8.7 Classification of Plutonic Rocks *135*
         Classification of Rocks Free of Feldspathoids *136*
         Classification of Silica Undersaturated and Ultramafic Rocks *139*

9 Sedimentary Rocks I: The Detrital Sedimentary Rocks *141*
   9.1 Weathering *141*
   9.2 Sedimentary Environments *142*

- 9.3 Stories the Rocks Can Tell *143*
    - Sedimentary Structures *144*
    - Allogenic and Authigenic Constituents *147*
    - Diagenesis *148*
- 9.4 Description and Classification of Clastic Sedimentary Rocks *148*
    - The Classification of Detrital Sedimentary Rocks *149*
- 9.5 Genetic Interpretations of Some Common Sandstones *152*
    - Quartz Arenite *153*
    - Arkose *153*
    - Lithic Arenites *154*
    - Other Sandstones *155*
- 9.6 Mudrocks *156*
    - Clay Mineralogy of Mudrocks *156*

10 Sedimentary Rocks II: The Chemical and Organic Sedimentary Rocks *157*

- 10.1 Chemical Composition of Seawater *157*
    - Order of Precipitation *159*
- 10.2 Limestone *159*
    - Biochemical Precipitation *161*
- 10.3 Dolostone *164*
- 10.4 The Evaporites *165*
    - Marine Evaporites *165*
    - Continental Evaporites *165*
- 10.5 Other Chemical and Organic Sediments *166*
- 10.6 Concretions, Nodules, and Armored Mudballs *169*
- 10.7 Classification of Chemical and Organic Sedimentary Rocks *170*

11 Metamorphism and Metamorphic Rocks *173*

- 11.1 Introduction *173*
- 11.2 Types of Metamorphism *173*
    - Regional Metamorphism *174*
    - Thermal Metamorphism *176*
    - Other Types of Metamorphism *178*
- 11.3 Pressures and Temperatures of Metamorphism *179*
- 11.4 Textures and Structures *180*
- 11.5 Classification of Metamorphic Rocks *181*
    - Foliated Rocks *181*
    - Nonfoliated Rocks *183*

Appendix I: The Geologic Time Scale   *187*

Appendix II: Elements of Crystallography   *189*

Appendix III: Bowen's Reaction Series   *193*

Appendix IV: Table of Elements   *195*

Glossary   *199*

References   *205*

Index   *209*

1
Minerals in the Past and Present World

Although Earth has often been referred to as the Blue Planet, its thin surface layer of air and water conceals a wealth of raw materials permitting people to mold and shape them into tools, machines, artwork, religious objects, and other practical and beautiful things limited only by our imagination and ingenuity.

Our attraction to, and utilization of, *minerals*, or natural inorganic solids of definite atomic structure and elemental composition, has a long and rich history. The earliest mineral commonly used by humans, of which we have evidence, was flint, or *chert*. Chert breaks to form a smoothly curving fracture surface like that of glass. More than 50,000 years ago, early peoples realized that this material could be chipped into suitable knives, spear points, and scrapers. The North American Clovis and Folsom cultures of 11,000 years ago are today known for their beautifully fluted points reflecting a high degree of skill and patience. Today, we have rediscovered this material and produce surgical blades of this and similar material.

Although chert is widespread and found on all the continents, some areas were inevitably graced with larger and/or higher quality deposits as well as the more highly valued *obsidian*, a dark glass of volcanic origin. Thus, somewhere, in a time before written history, the first extraction operations were begun. The material was collected or quarried, transported, and traded. This industry continues today on a global scale, utilizing a large number of mineral products. The minerals industry today generally involves four phases: exploration for a mineral or ore deposit; extraction of the raw material, usually in a developing country; manufacture of a finished product, usually in an industrial country; and finally, marketing of the finished product across the globe.

1.1 THE USE OF MINERALS BY EARLY CIVILIZATIONS

Although chert may have been the first mineral utilized by early peoples, the number of mineral types employed increased rapidly as people experimented with the effects of temperature and chemical processes and learned to control them to attain the desired effect. The human body's need for salt (*halite*) may have ensured that it was the first mineral discovered in deposits, although we have no records to suggest this. Today each American consumes about 10 pounds of salt per year. In the past, as well as in many countries today, blocks of salt were harvested and transported long distances from salt flats and playa lakes. The legendary camel caravans of Asia were begun to transport salt from the Middle East to points north and west. Salt was a valuable commodity because of its preservative properties, particularly in societies before the relatively recent invention of refrigeration as well as those in which such a convenience is a luxury item.

The native elements, particularly those that are metallic, were mined early in human history, and used mainly for jewelry and other artistic objects. The rarity of the native metals dictated that these objects were exclusively for use by royalty. Set into many of these gold, silver, or copper objects were many types of gemstones, including turquoise, emeralds, rubies, sapphires, and jade. Some of these minerals were carved into exquisite figurines with probable religious significance.

Still other minerals were utilized in various ways. Hematite and limonite were used as pigments. A variety of clays were used in making pigments, bricks, pottery vessels, and ornamental objects.

Eventually people learned to mine and extract the ores and to process them to extract the contained metal. Tin was one of the first metals extracted. When tin is combined with copper, the alloy known as bronze is formed. This mixture is much harder than the native copper alone, and thus, the world entered the Bronze Age. Another step was taken when humans learned to process iron-bearing minerals and make steel, an alloy much harder than bronze.

The earliest recorded attempts to understand the formation of minerals and their distribution are contained in the writings of Greek and Roman philosophers. Herodotus, Aristotle, and Pliny all attempted to describe and classify the occurrences of precious metals. However, the first real scientific monograph on the origin of ores was written by Georgius Agricola. Born in Saxony in 1494, Agricola was the first to realize the origin of hydrothermal veins (metals precipitated from hot fluids that flow through fractures in rock) and to recognize that they had formed after the rocks surrounding them had formed. This hypothesis was the first step in understanding the timing, origin, and development of ore deposits.

1.2 THE QUEST FOR ORE DEPOSITS

Throughout much of human history, we have had a fresh frontier to explore and exploit for minerals. Five hundred years ago, Christopher Columbus added the Americas to the list of territory explored by Europeans. For almost 500 years the moun-

FIGURE 1.1
View of an open-pit porphyry copper mine showing the benches in the pit and the network or roads spiraling down to the bottom of the pit.
(Photograph courtesy of Kennecott Corporation.)

tains of the Americas have been searched for mineral and rock deposits to fuel the industrial revolution.

We have been able to find rich *ore deposits*, economically valuable concentrations of minerals that can be profitably mined. However, in the last few decades, the quality or *grade* (quality of deposit usually expressed as a percentage, or in the case of precious metals, as ounces per ton of rock) of the deposits exploited has diminished and we have turned from mining relatively small but rich ore deposits to large deposits of lower grade. A classic example of this diminishing amount of quality deposits is the way copper (Cu) is mined now compared to 100 years ago. During those times in the copper mining camps of Arizona and New Mexico, the amount of ore mined exceeded 10 to 15 percent. Today in the same areas large deposits of disseminated copper, the so-called *porphyry copper* deposits, averaging less than 1 percent copper, are mined. The term *porphyry* refers to the presence of large crystals in the igneous rocks found in association with the deposits. Although the grade of copper in these deposits is low, it is compensated for by mining and processing large quantities of ore-bearing rock. As a result, everything about these contemporary copper mines is of grand scale (Figure 1.1): the diesel electric trucks hauling ore to the mill, the electric shovels, the large open pits, and even the size of the tailing piles are huge. Tailings are the crushed, processed rock left after extraction of the copper minerals.

1.3 ELEMENTAL ABUNDANCES IN THE EARTH'S CRUST

A cursory examination of the average composition of Earth's crust enables us to predict which mineral groups are most abundant in the crust (Table 1.1). Silicate minerals, composed predominantly of silicon and oxygen, are by far the most abundant minerals in Earth's crust. Thus, most minerals belong to a group known as the Silicates. Of the silicate minerals, the most abundant are those comprising the *feldspars* and the mineral *quartz* (SiO_2). *Feldspars* are aluminum-bearing silicates of calcium, sodium, and potassium. Other abundant mineral groups include the carbonates (e.g., $CaCO_3$) and oxides, reflecting the overwhelming amount of oxygen in the Earth's crust. Note that the majority of the most prized metallic elements, or those that are the basis for industrial civilization—with the exception of aluminum and iron—form an extremely small fraction of the crust. The search for these rarer elements, including copper (Cu), molybdenum (Mo), silver (Ag), gold (Au), lead (Pb), and zinc (Zn), among others, has historically employed geologists having an understanding of the special conditions that promote metallic ore deposit formation.

The metals have historically occupied a special place in the hearts and minds of civilizations. Metals have versatility; they are essential for the construction of machines that replace muscle power in much of the industrialized world. Probably

TABLE 1.1
Average Chemical Composition of the Earth's Crust, in weight percent

Element	Percent of Crust by Weight
Silicon	27.7
Oxygen	46.6
Aluminum	8.1
Magnesium	2.1
Calcium	3.6
Iron	5.0
Sodium	2.8
Potassium	2.6
All others	1.5

(Source: Parker, 1967, Composition of the Earth's Crust, U. S. Geol. Surv. Prof. Paper, Table 18.)

the most versatile is the alloy (manufactured mixture of two or more metals) metal *steel* in its various forms. Some varieties are *stainless*, meaning that they are highly resistant to *oxidation* (rusting); others are elastic; and some are exceptionally strong or resistant to melting at high temperatures. These properties are achieved by the addition of small amounts of alloy metals such as manganese (Mn), chromium (Cr), or molybdenum (Mo).

The persistent work of organic chemists has been rewarded; there is now a whole new family of substances composed of petroleum-based synthetic compounds that is proving just as versatile as the metals—the plastics. Like the metal alloys, however, plastics also require a wide range of minerals used as fillers that give them special properties such as hardness and density. Thus, the need for minerals and the ways in which they are used will continue to grow.

2
The Chemical and Physical Properties of Minerals

Minerals are made up of chemical elements. There are 92 naturally occurring elements (Figure 2.1). The element with *atomic number* 92 is uranium (U). There are 12 other so-called transuranic or artificially produced elements known, such as lawrencium (Lw), mendelevium (Md), and nobelium (No), which usually form in extremely small quantities in nuclear reactors and atom smashers. In this chapter we will explore the properties of elements and how they can be combined to form compounds.

2.1 STRUCTURE OF THE ATOM

The smallest part of an element that still retains all the element's properties is an *atom*. An atom is composed of a number of subatomic particles, the three essential ones being

protons (+ charge)
neutrons (no electrical charge)
and, *electrons* (− charge).

Protons and neutrons occupy the central region of the atom called the *nucleus*. The mass of a proton is approximately equal to that of a neutron, which is much greater than the mass of an electron. Because neutrons and protons are densely clustered within the nucleus, most of an atom's mass is concentrated there. One of the mysteries of physics is how protons with a similar positive charge can be so tightly clus-

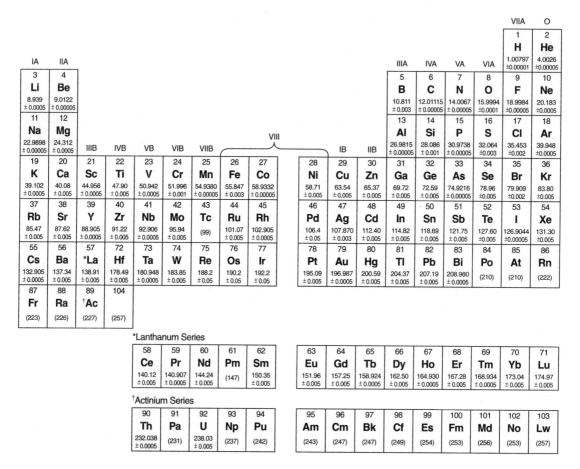

FIGURE 2.1
The Periodic Table. Elements with similar chemical properties are arranged in columns.

tered in the nucleus. There must be a very strong force, the *nuclear force*, binding them together. Electrons, which possess a negative electrical charge, encircle the nucleus, as if in orbit. An atom could be described as a fairly small nucleus surrounded by a cloud of electrons.

An element, and thus its properties, is defined by the number of protons in the nucleus. This is also the element's *atomic number*. Thus, the number of protons is unique to each element (Figure 2.1). For example, hydrogen (H) has one proton, silicon (Si) has 14, gold has 79, and so on. The *atomic weight* of an element equals the number of protons plus neutrons. Thus, hydrogen, having an atomic weight of one, has no neutrons, but carbon (C) with an atomic number of six, and atomic weight of 12, contains six neutrons in the nucleus. The weight of an electron is 1/1837 that of a proton, emphasizing the fact that most of an atom's weight is concentrated in the nucleus.

2.2 IONS AND ISOTOPES

An alphabetical list of the elements and their symbols, atomic numbers, and atomic weights is given in Appendix IV. Careful comparison of atomic number and atomic weight reveals that there appears to be no systematic number of neutrons. Hydrogen has none, helium (He) and carbon (C) have the same number of neutrons and protons, and other elements [e.g., strontium (Sr) and lead (Pb)] have more neutrons than protons. In fact, some elements have a variable number of neutrons in the nucleus. Elements with different numbers of neutrons are called *isotopes*. An example of an element having several isotopes is strontium, which has an atomic number of 38. The three isotopes of Sr and the makeup of the nucleus is shown here:

$$^{86}Sr \qquad ^{87}Sr \qquad ^{88}Sr$$
$$p = 38, n = 48 \qquad p = 38, n = 49 \qquad p = 38, n = 50$$

The average atomic weight of strontium (87.62) is the average of these three isotopes weighted to their relative abundances. The average atomic weight of 87.62 indicates that ^{88}Sr is the most abundant isotope of Sr. Isotopes are important in tracing and delineating many geologic processes.

Individual isotopes possess a unique combination of electrical charge and mass. That individuality makes it possible to mechanically separate isotopes and determine the relative amounts of each. Isotope ratios are used to determine past temperatures of oceans and lakes, the movement of atmospheric and groundwater, and the ages of minerals, rocks, and the Earth.

The high energy necessary to alter the makeup of the nucleus of *stable* isotopes is produced only in atom smashers, nuclear reactors, or nuclear explosions, where neutrons or other subatomic particles are produced in great quantities or accelerated to very high speeds and made to collide with atomic nuclei. These collisions knock protons and/or neutrons out of the nucleus, thus releasing large amounts of energy in the form of heat and radiation.

Some naturally occurring isotopes are *radioactive*; they naturally and spontaneously change their nuclei compositions. Some of the most common of these are ^{40}K (potassium), ^{238}U, ^{235}U, ^{232}Th (thorium), and ^{87}Rb (rubidium). Because the rate of decay (change) of these isotopes is not affected by changes in heat or pressure, and because the decay rates are known and the relative amounts of parent and daughter elements can be determined, they may be used to date many geologic events. Most of the problems associated with radiometric dating are the result of leakage of either the parent or daughter element. However, these problems are usually recognizable and often used to the advantage of geological research.

Ions are atoms that are charged due to a deficiency or overabundance of electrons. An electrically neutral atom has an equal number of electrons and protons. Ions form by the addition or removal of electrons. This fact illustrates the difference between ordinary chemical processes and nuclear processes. However, *nuclear* processes involve the nucleus and, thus, large amounts of energy are required to break the nuclear force tightly binding protons and neutrons.

In contrast, ordinary chemical reactions involve the transfer or sharing of electrons, usually only one or two of the outermost electrons. The two types of ions are *anions*, which are negatively charged, and *cations*, which are positively charged. Examples of some common cations are sodium (Na^+), K^+, magnesium (Mg^{2+}), iron (Fe^{2+}) and Ca^{2+}. Examples of some common anions are carbonate ($CO_3^=$), sulfate ($SO_4^=$), oxide ($O^=$), and chloride (Cl^-). Cations and anions combine to form *compounds*. For example, calcium and carbonate combine in equal proportions to form the compound known as the mineral calcite ($CaCO_3$). Most minerals are compounds, and most of them are complex compounds, involving several elements.

About 3,000 minerals have been identified and described. An average of five to 10 new minerals are discovered and described every year. Only about five percent of the known minerals are common, and about 10 minerals constitute more than 90 percent of Earth's crust.

2.3 CHEMICAL BONDING

As a result of experiments, we know that bonds are electrical in nature. In order to understand the role of electrons in the bonding process, let us examine the structure of an atom. Figure 2.2 shows a schematic representation of an argon (Ar) atom. Chemists have determined that the most stable configuration of electrons is two in the inner shell, eight in the next shell, eight in the third shell, and so on. Argon, with an atomic number of 18, has a stable configuration of electrons (2 + 8 + 8 = 18), distributed among its first three shells, making it an electrically neutral element. In this configuration it is unlikely to lose or gain electrons, making argon a relatively inert element.

FIGURE 2.2
Schematic representation of an argon atom, showing the nucleus at the center surrounded by electrons in different shells.

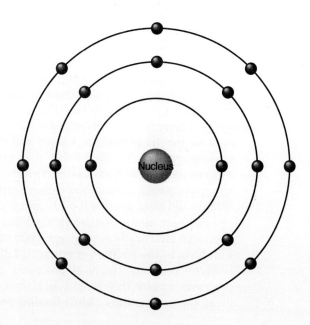

The arrangement of the *Periodic Table* (Figure 2.1), was devised by a Russian chemist, Dmitri Mendeleev. It was first published in 1869 when only 63 elements had been discovered and described. Mendeleev left gaps in his table that were subsequently filled in as the elements were discovered. In the Periodic Table, elements having similar properties are arranged in columns (groups). For example, the noble gases, including argon, are considered relatively inert. The reason for this is the number of electrons and their arrangement among the electron shells as shown below:

helium	2 electrons
neon (Ne)	2 + 8 electrons
argon	2 + 8 + 8 electrons
krypton (Kr)	2 + 8 + 18 + 8 electrons
xenon (Xe)	2 + 8 + 18 + 18 + 8 electrons

In noble gases all the lower energy levels are filled to capacity. Except for helium, the highest energy level contains eight electrons—an electron arrangement that is most stable. In Group IA all of the elements have one electron in the highest energy level. The Group IIA elements have two electrons in the highest energy level. For example:

Sodium, atomic number 11, has 2 + 8 + 1 electrons
Potassium, atomic number 19, has 2 + 8 + 8 + 1 electrons
and,
Magnesium, atomic number 12, has 2 + 8 + 2 electrons
Calcium, atomic number 20, has 2 + 8 + 8 + 2 electrons

Group VIIA elements, with the exception of hydrogen, have seven electrons in the outer shell. An example is chlorine, with an atomic number of 17. Its electron distribution from inner to outer shells is two, eight, and seven. The grouping of elements in the Periodic Table allows the chemist or geochemist to predict the behavior of elements in such processes as substitution, reactivity, and bonding.

Four types of chemical bonding are recognized: ionic, covalent, metallic, and Van der Waals. *Ionic bonding* is the bonding produced by the mutual attraction of positively and negatively charged particles and is best illustrated by the structure of the mineral halite, NaCl. In this structure, positive ions of Na alternate with negatively charged ions of Cl (Figure 2.3). The attraction between unlike electrical charges holds the structure together and in a three-dimensional diagram one can see that each Cl^- ion is surrounded by six Na^+ ions, and each Na^+ ion has six Cl^- ions as its closest neighbors.

Covalent bonding involves a sharing of electrons rather than a transfer of electrons. Let's look at the chlorine atom again. It has an atomic number of 17 with seven electrons in its outer shell. At room temperature, chlorine exists as a pure element as a gas, Cl_2. Why is it a diatomic molecule? A look at Figure 2.4 shows why this is a stable arrangement. Two electrons are shared in the outer shell so that both atoms have eight, a stable configuration. This is an example of a covalent bond. Covalent bonds are directional and much stronger than ionic bonds and very common in min-

FIGURE 2.3
The atomic structure of halite.
Source: Klein, C., and C. S. Hurlbut. 1985. *Manual of Mineralogy*. 20th ed., New York. John Wiley & Sons. Fig. 9.34, p. 321.

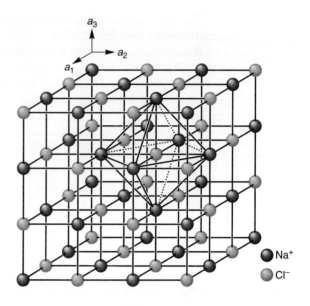

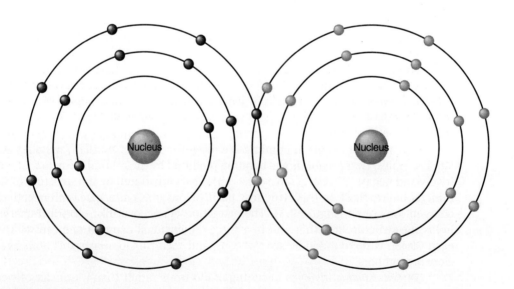

FIGURE 2.4
Schematic representation of a diatomic molecule of chlorine, illustrating the sharing of electrons.

erals. The bond between silicon and oxygen in the silicate minerals is a covalent bond, forming a SiO_4 tetrahedron.

Metallic bonding is found in the native metals—gold, silver, platinum (Pt), and copper. It involves the sharing of electrons but the electrons are free to move around. This property produces high electrical conductivity common in metals as well as other properties. This bond is also nondirectional, producing crystal structures of high symmetry.

The *Van der Waals bond* is a very weak bond produced by the residual electrical fields existing among essentially neutral atoms. In the mineral *graphite*, loosely bonded sheets of carbon atoms are formed of covalently and metallic-bonded carbon atoms. However, the attachment between adjacent sheets by the weak Van der Waals bonds causes the mineral to be soft and break into parallel sheets.

2.4 INTRODUCTION TO THE REALM OF MINERALS

Let us begin our discussion of minerals by talking about the difference between different minerals and the difference between minerals and rocks. A *mineral* is defined as a naturally occurring element or compound having an internal ordered structure and a definite chemical composition. Let's briefly review the essential parts of this definition. That a mineral is naturally occurring means that it occurs in nature—not manufactured, nor an artifact. Compounds produced by industrial processes are not considered minerals. Second, minerals have an internal ordered structure. This internal ordered structure, together with the atoms or ions included in that structure, is unique for that mineral. Two minerals may have the same ordered structure, such as the minerals *halite* and *galena*, but the chemical components of those structures are different for the two minerals. Third, each mineral has a definite chemical composition. For some minerals, such as quartz (SiO_2), the given chemical composition is very close to the actual chemical composition. Such minerals are said to be *stoichiometric*. Other minerals such as *olivine* and the feldspars have a chemical composition that varies within a specified range. For example, olivine has a composition that ranges between Mg_2SiO_4 and Fe_2SiO_4. It is impossible to determine the chemical composition of such compositionally variable minerals in hand-specimen; instead, chemical analysis or sophisticated microscopy methods are required.

In summary, the characteristics making one mineral different from another are combinations of chemical composition and internal ordered structure. Minerals that possess an identical chemical composition but different structures are known as *polymorphs*. Examples of polymorphism are the minerals *aragonite* and *calcite*. The composition of both is $CaCO_3$, but their internal structure, and thus their crystal form and other physical properties, differs. Another example of a polymorph pair are graphite and diamond, both composed of carbon. Some minerals mimic others, both maintaining an identical external crystal form. Through weathering or alteration the original mineral may be transformed into another. Such minerals are called *pseudomorphs* (Figure 2.5).

FIGURE 2.5
Pseudomorphs of goethite after pyrite, Chaffee County, Colorado. The original pyrite, crystallized as cubes, has been replaced by goethite but retains the original external cubic form. Cubes are 4 to 10 mm on a side.

A *rock* is defined as an aggregate of the crystals or particles of one or more minerals. A few rocks are composed of only one mineral. Such rocks are said to be *monomineralic*. An example of a monomineralic rock is the sedimentary rock (Chapter 9) *limestone*, composed of the mineral *calcite*. Most rocks, however, contain more than one mineral. An example is the igneous rock (Chapter 8) *granite*, which contains three essential minerals—quartz, alkali (aluminum-potassium) feldspar, and plagioclase feldspar.

2.5 PHYSICAL PROPERTIES OF MINERALS

A number of physical properties of minerals can be used to identify all common minerals with a high degree of certainty without damaging the specimen under study. These properties include crystal form and habit, color, cleavage and fracture, hardness, luster, specific gravity, streak, and other more esoteric properties such as fluorescence, magnetism, and taste.

Color

Mineral color is one of the most obvious, easily recognized properties of minerals. However, it is not a useful property for most minerals. Only a few minerals such as gold, sulfur (canary yellow), pyrite (brassy yellow), and amethyst (purple) have a consistent color and thus, in many other instances, additional properties must be evaluated to distinguish among similarly colored minerals.

Crystal Form and Habit

Crystal form is a useful diagnostic property. However, many specimens do not exhibit a crystal form, either because they are a fragment of a larger crystal or very fine grained (mineral grains ≤ 1 mm diameter). However, some minerals commonly form good crystals. Among these are quartz, tourmaline, corundum, garnet (Figure 2.6a–d), and quite a number of other common minerals (halite, galena, pyrite, and fluorite) that occur in cubic form. The student must take care to distinguish between crystal faces resulting from crystal growth and *cleavage faces* resulting from breakage along zones of weakness in the crystal. Cleavage will be discussed in greater detail in the next section.

Hâbit refers to the general shape of the crystals or to the arrangement of a group of crystals. It is a less rigorous description than is crystal form. For example, kyanite typically forms bladed crystals, pectolite crystals are acicular (needlelike), and barite crystals are tabular (Figure 2.7a–c).

Cleavage and Fracture

Mineral cleavage is a difficult concept to grasp upon introduction. *Cleavage* can be defined as the tendency of a mineral to break along predefined planes of weakness. These planes are a manifestation of the internal structure of the crystal in which cleavage planes, or planes of potential fracture, occur where bonds are at their weakest. Not all minerals possess cleavage, but only crystalline solids can have cleavage. Volcanic glass does not have cleavage because it is not a crystalline solid. If broken, glass breaks with an uneven, but smoothly undulating fracture surface called *conchoidal fracture*. Quartz, which possesses an internal ordered structure and commonly forms crystals, also does not cleave. Quartz also breaks with conchoidal fracture. Fluorite crystallizes as cubes, but cleaves into octahedrons, or eight-sided solids. In contrast, halite and galena tend to grow as cubes and also cleave into cubes.

In analyzing the cleavage of an unknown mineral it is important to determine how many cleavage planes are present and what the angular relationships are among the planes. It may sometimes be important to determine how readily a plane(s) cleaves in each "direction" or set of parallel surfaces. For example, the mica group (muscovite, biotite, etc.) has perfect cleavage in one direction. This cleavage is the major diagnostic property of micas. Halite and galena possess cubic cleavage—perfect cleavage in three directions at right angles to each other. Calcite has rhombohedral cleavage—perfect cleavage in three directions—but not at right angles. Fluorite possesses octahedral cleavage—perfect cleavage in four directions. Note that each direction is repeated so that cubic cleavage with three directions of cleavage forms six-sided cubes; rhombohedrons also have six sides, but not at right angles; and octahedrons, the result of octahedral cleavage, have eight sides.

FIGURE 2.6
Crystals of: a) quartz (upper left, 3 cm long); b) corundum (upper right, 2 cm high); c) garnet (lower left, 1.5 cm in diameter); and, d) fluorite (lower right, 2.5 cm cubic crystals).

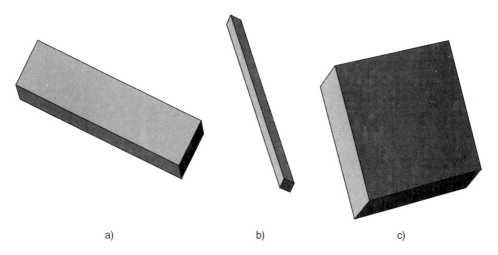

FIGURE 2.7
Examples of: a) bladed; b) acicular; and, c) tabular crystal forms.

Hardness

Hardness is a measure of a mineral's resistance to scratching. Some students interpret hardness to mean resistance to pulverization, resulting in laboratory specimens soon reduced to piles of powder. A scale of *relative* hardness, the *Mohs Hardness Scale* (Table 2.1), is useful for comparing and determining the hardness of a mineral. Two common minerals easily confused by students unfamiliar with the concept of hardness are calcite and quartz. A simple hardness test with a glass plate or a knife blade serves to distinguish the two. Similarly, two other commonly confused minerals, purple fluorite (H = 4) and amethyst (H = 7), are easily distinguished by hardness. Amethyst easily scratches a glass plate, but fluorite does not.

TABLE 2.1
Moh's Scale of Hardness

Rank	Mineral	Test of Hardness
1	Talc	
2	Gypsum	Fingernail is about 2.5
3	Calcite	
4	Fluorite	
5	Apatite	Nail is about 5.0
6	Orthoclase	Scratches glass with difficulty
7	Quartz	Scratches glass easily
8	Topaz	
9	Corundum	
10	Diamond	Cuts glass easily

Luster

Luster is the surface appearance of a mineral's reflected light. The two basic types of luster are *metallic* and *nonmetallic*. Metallic luster is the most easily recognized—it has the familiar appearance of a metal. The sulfides and the native elements such as silver and gold have metallic lusters. Nonmetallic lusters are of several types, the most common of which is *vitreous* or glassy luster. Examples of minerals with glassy luster include calcite, quartz, and micas. Other minerals possess an *earthy* luster, characterized by a dull appearance. Some varieties of hematite and many clay minerals exhibit this type of luster. Some minerals, such as satin-spar gypsum, possess a *pearly* or satiny/silky luster, while others such as sphalerite possess a *resinous* (greasy) luster. A few minerals, such as zircon and rutile, have a diamond-like *adamantine* luster.

Specific Gravity

The density of a mineral is a measure of its weight per unit volume expressed as grams per cubic centimeter (g/cc). *Specific gravity* is the ratio of the weight of a given substance to the weight of an equal volume of water. An average specific gravity is in the range of 2.7 to 3.0 g/cc. For example quartz and feldspars have a specific gravity of 2.7, while calcite has a specific gravity of 3.0 g/cc. Most minerals with metallic luster have specific gravities of 4.0 or more. For example, galena has a specific gravity of 7.5 g/cc, and that of native gold ranges from 14.0 to 19.0, depending on the quantity and nature of other alloyed elements. A few minerals with vitreous luster also have above-average specific gravities that are diagnostic. Such minerals include fluorite (3.2), celestite (4.0), and barite (4.4).

The Jolly balance has been designed to rapidly and easily calculate the specific gravity of an unknown mineral. However, with a little bit of practice hefting specimens of various minerals of known specific gravity and of various sizes, the student can quickly acquire sufficient skill to differentiate between various minerals on the basis of specific gravity.

Streak

The *streak* of a mineral is the color of its powder. The streak is most easily determined by rubbing the mineral against an unglazed tile of white porcelain and examining the color of the resulting powder. Most minerals have streaks very similar to that of their colors. Thus, this test is most useful for differentiating between minerals such as magnetite and specular (sparkly) hematite, with similar surficial colors. Magnetite has a black streak, the same color as the mineral, but hematite has a brick-red streak.

Other Physical Properties

A large number of other physical properties may be useful or diagnostic for a few minerals or mineral groups. These include taste (sylvite and halite), magnetic properties (magnetite and pyrrhotite), electrical properties (quartz), effervescence (bubbling chemical reaction) in acids (the carbonate minerals), radioactivity (uranium and thorium-bearing minerals), fluorescence in ultraviolet light (many minerals), as well as a wide range of optical properties in the realm of optical mineralogy. A note of caution should be made here—some minerals contain toxic or radioactive elements and one *should not* indiscriminately *taste* unknown minerals.

2.6 CHEMICAL CLASSIFICATION OF MINERALS

All minerals can be classified in various ways. Most systematic mineralogy books use a geochemical classification as the basis for organizing minerals. In addition, minerals of each group are categorized according to similarity of crystal form. A standard chemical classification of minerals is as follows:

1. Native Elements—Au, Ag, Pt, Cu, S, and C.
2. Sulfides and Sulfosalts—also called the ore minerals, include the economically most important minerals.
3. Oxides and Hydroxides—includes the source of the metals titanium, iron, and aluminum.
4. Halides—minerals whose anions include bromine, iodine, fluorine, and chlorine.
5. Carbonates, sulfates, nitrates, borates, chromates, molybdates, phosphates, arsenates, and vanadates include a variety of economically important minerals used in the manufacture of fertilizers, cement, plastics, ceramics, and glass, as well as fiberglass.
6. Silicates—or rock-forming minerals. Within this large group are included minerals used in glass and ceramics, as refractory materials, as gemstones, and as ore minerals for elements used in defense and high-technology industries such as beryllium, lithium, and the rare-earth elements.

3
The Ore Minerals:
Native Elements, Sulfides, and Sulfosalts

These groups of minerals are called ore minerals because almost all are of economic importance (Table 3.1). The majority of these minerals are a source of the metallic elements constituting the cation needed to form the minerals and deposits rich enough to be economically extracted.

The native elements form a rather small group of minerals; however, all minerals in this group are of strategic or economic importance because of their rarity or special properties. Based on their physical properties, the native elements can be subdivided into two groups, the metals and the nonmetals. The metallic group includes the precious metals gold, silver, and platinum, as well as the industrially important metal, copper. The nonmetallic group exhibits diverse physical properties and includes sulfur, an essential and widely used industrial mineral, and the two forms of carbon: graphite and diamond.

3.1 THE METALLIC NATIVE ELEMENTS

As a group, the metallic native elements are distinguished by their high specific gravity (5.0 to 21.0 g/cc), metallic luster on a fresh surface, softness, and malleable character (Table 3.1). Because these minerals are *malleable*—meaning that they can be pounded into shape—and because they occur in the native state (and therefore need not be refined), they were among the earliest metals used.

Silver

The symbol for the element silver, Ag, is derived from the latin word for silver, *argentum*. Although found in the native state in irregular masses or as coarse or fine wires (Figure 3.1) in hydrothermal veins, most of the world's supply of silver is mined in the form of sulfides. Most native silver is dark gray or black because it is easily oxidized, or tarnished upon contact with air.

TABLE 3.1
The Native Elements

Mineral Name	Chemical Composition	Hardness	Specific Gravity	Uses	Diagnostic Properties
Silver	Ag	2-3	10.0-12.0	Photographic Emulsions, Jewelry	Metallic silver color tarnishes, malleable, ductile.
Gold	Au	2-3	19.0	Reflector, Conductor, Jewelry	Metallic gold color, does not tarnish, malleable, ductile.
Copper	Cu	2-3	9.0	Conductor	Metallic copper color, tarnishes green, malleable.
Platinum	Pt	4	15.0-21.0	Jewelry	Metallic silver color, very high specific gravity, malleable
Iron	Fe (with Ni)	4	7.0-8.0		Metallic gray. Occurs in meteorites alloyed with Ni.
Sulfur	S	2	2.0	Used in manufacture of rubber	Canary yellow color, burns easily, low specific gravity.
Graphite	C	1	2.0	Refractory crucibles	Black submetallic luster, greasy feel, low hardness and specific gravity.
Diamond	C	10	3.5	Abrasive, gemstone	Variable color, octahedral crystals, exceptional hardness.

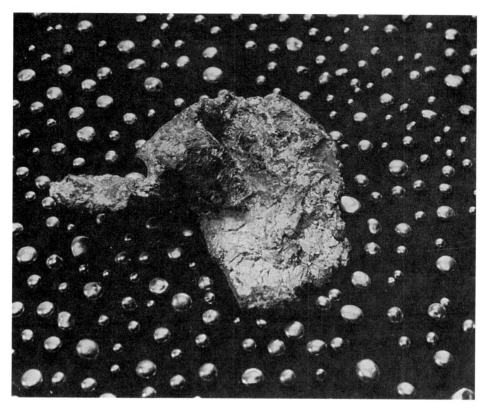

FIGURE 3.1
Native wire silver from Chile.
Source: Hurlbut, C. S. (1970). *Minerals and Man*. New York: Random House, p. 152.

The Batopilas District Legend has it that the great silver mining district of Batopilas in western Chihuahua's Barranca del Cobre (Copper Canyon) region of the Sierra Madre Occidental, Mexico, was discovered in 1632 by a small group of Spaniards who had come up the Rio Batopilas, a tributary to the Rio Fuerte. A large mass of native silver reflected sunlight, creating a bright luster which shined into the Spaniards' eyes. The mass of native silver remained untarnished due to the abrasion of sand grains carried in the stream at the base of a 5,000-foot-deep canyon. The Batopilas mining district is famous for its large quantities of native silver. Irregular masses are randomly distributed in veins in igneous rock, as wire silver, and as skeletal featherlike crystals called *herringbone*. The larger masses of silver discovered were too large to be carried out of the mine. The malleable and ductile nature of the native metal was not susceptible to explosives. The challenge of breaking them up into pieces small enough to be carried out on the backs of miners was accomplished by

FIGURE 3.2
View of Creede, Colorado, at the northern edge of the Creede Caldera. The mountain in the middle of the photograph forms the resurgent dome in the caldera's center. Creede is at lower left.

the use of axes and handsaws. Even today, panning for silver in the gravel stream bed of the Rio Batopilas yields numerous specimens of wire silver.

The Creede District, Colorado   Another famous silver mining district is the Creede District in the San Juan Mountains of Colorado. The Creede District sits at the edge of the Creede Caldera, a large circular depression produced by the collapse of the roof of a magma chamber. Numerous mines in the district, situated mostly to the north of the town of Creede (Figure 3.2), produce masses of silver and wire silver. The mineralization at Creede is due to the proximity of intrusive igneous rocks that provided heat and metallic elements, and the older intruded and pervasively fractured rocks. This fractured bedrock allowed metal-rich solutions to circulate and precipitate silver and other minerals within.

The history of the Creede mining district is similar to that of many of the boom and bust mining towns so prevalent in the Colorado Rockies. Silver was discovered in the Creede area in 1889 and the population of the town of Creede grew to 10,000 by 1892. Shortly thereafter, in 1891, the Amethyst Vein—a major zone of mineralization—was discovered and actively mined until 1930. However, in 1893 the Silver Act dropped its price from $1.29 to 50 cents per pound, and the Creede silver boom lost a lot of its "luster." More recently, however, the United States Geological Survey (USGS) has begun an intensive study of the Creede ore system in an effort to understand the process of ore formation. The benefits derived from this endeavor may well be greater than the monies gained by exploiting the silver in this district.

Uses of silver are many, including electronics equipment, silverware, and photographic emulsions, its major use. Mexico has long been the world's leading producer of silver. The Spaniards were quick to exploit its incredible silver riches. Silver was first discovered in Mexico at Taxco about 1550 A.D., in the state of Guerrero, south of Mexico City. In the 26 years following this initial discovery, the rich silver deposits of Guanajuato, Zacatecas, and Pachuca were discovered. In the United States the leading silver-producing states are Idaho and Nevada.

Gold

Native gold occurs in a variety of forms—as poorly formed octahedral crystals, and as dendritic or reticulated masses that reflect its cubic internal arrangement. Gold is commonly found as nuggets (Figure 3.3) and gold dust, which represent the effects of abrasion in a stream or wave-dominated environment, or precipitation induced by bacteria. Most gold contains some silver, but it may also contain small amounts of copper and iron. The presence of these lighter metals reduces the specific gravity of native gold from 19.3 to about 15.0 g/cc. An alloy of gold and silver is called *electrum*. The primary occurrence of gold is in hydrothermal (hot fluid) deposits. These are formed as two types: in quartz-rich veins, the so-called *lode deposits*, and in *disseminated* deposits as microscopic particles distributed throughout the rock.

Secondarily, gold occurs in *placer* deposits which represent the weathering products of lode deposits. Because gold does not oxidize or weather to other minerals, it persists in the stream environment and its high specific gravity allows its easy separation from the much lighter quartz and feldspar sand and gravel. This is the principle of gold-panning or hydraulic mining used to the forty–niner's advantage in the Sierra Nevada of California, and in Alaska, Colorado, and throughout the world.

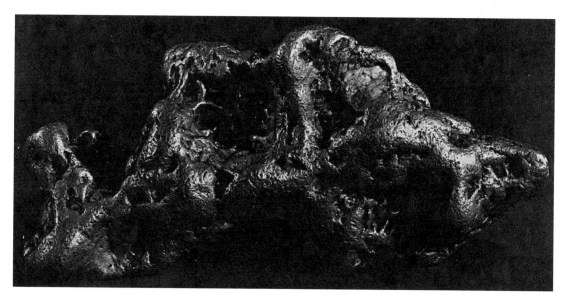

FIGURE 3.3
Gold nugget from California
Source: Smithsonian Institution Photograph 77-9080.

The Witwatersrand Basin In 1886 two prospectors, George Harrison and George Walker, walked upon an outcrop on the southern slopes of the Witwatersrand of South Africa and discovered what was to become the greatest gold mining district in the world. The deposit seems to be a paleoplacer, or ancient placer deposit, now consolidated into a *conglomerate*, a sedimentary rock composed of cemented, rounded cobbles and pebbles. However, much debate persists about the origin of the gold in the Witwatersrand. The average grade of the gold in the basin is about 0.3 oz./ton, and more than 1.2 billion ounces of gold have been recovered. Since 1886, the "rand" has produced over 50 percent of the world's gold.

The ore-bearing zones are in a Precambrian sequence (2.8–2.7 billion years old) of sedimentary rocks including quartzites, conglomerates, iron formations, and clay layers. The gold-bearing conglomerate zones are locally called *reefs*—an antiquated mining term for an ore-bearing horizon. The deepest mines in the district are some of the deepest mines in the world, reaching depths of more than 10,000 ft and having temperatures of up to 120°F.

The price of gold is an interesting story itself. Undoubtedly it had been prized since long before the time of Christ for its rarity and physical properties. The tombs of the Egyptian Pharoahs contained a variety of gold artifacts. For many years the United States used a gold monetary standard, set the world price of gold, and prohibited its citizens from owning gold for uses other than as jewelry. From the 1930s into the mid-1960s the price of gold was set at $35.00/troy oz. In 1968, the U.S. government allowed the price of gold to seek its own level. By 1980, it had soared to more than $600/oz. before dropping to more moderate levels at around $300–$400/oz.

The uses of gold reflect its unique properties. Gold can be pounded into foil only a few atoms thick, pulled out into a thin wire, or coated onto a surface only a few microns thick. Its reflectivity and resistance to tarnishing make it an ideal reflector coating. The bulk of gold produced is used in jewelry or bullion and coinage for use as a hedge against a disproportionate amount of paper currencies in circulation.

Copper

Most copper used for industry and other purposes is mined in the form of sulfides. A very small fraction of all copper mined occurs in its native state, usually in irregular masses, dendritic plates and scales (Figure 3.4), and, in some instances, in groups of deformed crystals—usually octahedrons, cubes, or dodecahedrons.

Copper tarnishes easily to a green crust. It is widely used in electrical wiring, and in much smaller amounts in jewelry and as an alloy in combination with beryllium (Be). The most famous occurrence of native copper in North America is probably that on the Keweenaw Peninsula of Michigan, on Lake Superior's south shore. In this area, basaltic lava flows of late Precambrian age (1.0–1.6 billion years of age) contain native copper mineralization in cavities in the flows and as veins cutting the flows. Because only very low grade ore remains in these deposits, most of the copper mined in the United States now comes from large low-grade deposits in New Mexico and Arizona. These *porphyry copper* deposits, so called due to the porphyritic

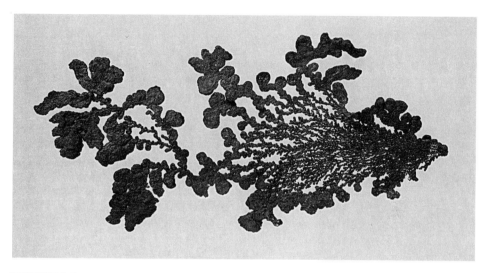

FIGURE 3.4
Native copper that has crystallized in a dendritic pattern.
Source: Smithsonian Institution Photograph 36474-C.

texture (large crystals scattered in a fine-grained matrix) of the igneous rock in which the copper is found, contain small amounts of native copper. Spaniards visiting the area of the Santa Rita porphyry copper deposit in southwestern New Mexico noted that the Native American inhabitants had exploited the native copper deposits there.

Platinum

Platinum is a silver-colored metal with the highest specific gravity—21.4 g/cc—of any of the native metals in its pure state. Most native platinum, however, is alloyed with other platinum group metals such as iridium (Ir), osmium (Os), rhenium (Rh) or palladium (Pd) or with Fe and Cu, reducing its specific gravity to a range of 14.0 to 19.0 g/cc. Platinum occurs as minute grains disseminated in the ultramafic portions of *stratiform complexes*. These complexes originated as injections of basaltic magma that differentiated during crystallization into a variety of rock types. The lower portions of these bodies contain rocks rich in magnesium and iron, hence the term, *ultramafic*. These rocks are composed mostly of pyroxenes and olivine with minor amounts of plagioclase and oxides. Stratiform complexes are widely distributed in the geologic record and occur on all of the continents; however, the largest reserves of platinum are contained in the Merensky Reef of the Bushveld Complex in South Africa. This "reef," in the lower part of the complex, is a layer about one-foot thick, extending for many miles. It has a platinum content of about one-half ounce per ton of rock.

The best known stratiform complex in the United States is the Stillwater Complex of southern Montana. It has been intermittently mined for chromium since World War II, and is being explored for platinum mineralization.

Because of its high melting point of 1755°C, platinum is used in high-technology and defense instruments and electrical equipment, and as a catalyst. It is also used in jewelry.

3.2 NONMETALLIC NATIVE ELEMENTS

Sulfur

The primary source of sulfur production is salt domes and volcanoes. Sulfur is found in the caprock, or the rock overlying most salt domes. Caprock is composed of calcite, anhydrite, gypsum and locally, sulfur. Of 230 salt domes on the Gulf Coast of Texas and Louisiana, only 24 have produced sulfur.

Apparently, a shallow caprock depth, the presence of petroleum, and relatively low temperatures must be present for native sulfur to form. These conditions allow the presence of *Desulfovibrio desulfuricans* bacteria. The bacteria use the petroleum as an energy source in a reaction that liberates the sulfur in the anhydrite ($CaSO_4$) and gypsum ($CaSO_4 \cdot H_2O$) as hydrogen sulfide (H_2S). The hydrogen sulfide is subsequently oxidized to native sulfur.

The extraction of sulfur from the salt dome caprock utilizes its low melting point (175°F) and is known as the Frasch Process. In this procedure, a well is drilled into the sulfur deposit and hot water (225°F) is piped in to melt the sulfur. The molten sulfur is then pumped or forced by compressed air to the surface, where it is pumped to a holding pond and solidifies. Alternatively, the molten sulfur may be transported in railroad tank cars.

Sulfur is widely used in the chemical industry, mainly in the form of sulfuric acid. Elemental sulfur is used in explosives, rubber, plastics, and pharmaceuticals. Increasingly, as environmental regulations become more strict, more sulfur will be produced as a by-product of scrubbing processes used to remove sulfur from stack gases at refineries and smelters.

Diamond and Graphite

Diamond and graphite are polymorphs (variants of crystal form) of the element carbon, but their physical properties lie at opposite ends of several spectra. While diamond is the hardest known natural substance, graphite is one of the softest. On one hand, the best specimens of diamond, when cut to advantage, possess a transparency and brilliance that have inspired legends, while graphite comes in one color, black. Diamonds occur in a wide range of colors ranging from colorless to yellow, blue, green, and red, due to minor impurities. The most prized stones are those which are either yellow (the Tiffany Diamond) or bluish white (the Hope Diamond).

Diamonds occur primarily in a rock known as *kimberlite*. The kimberlite occurs in funnel-shaped deposits called kimberlite pipes. The rock is typically weathered and oxidized to form the so-called "yellow ground" at the surface. Mining of diamonds in these pipes has produced some of the deepest open pit mines in the world. Diamonds in this rock type may be *euhedral*, a term indicating the presence of well-formed crystal faces. Such crystals are commonly octahedrons, usually with curved faces. The euhedral nature of the diamonds suggest that they crystallized from a melt, thus allowing determination of a minimum depth of formation. Because diamond is stable only at pressures above 30 kilobars, and at the temperatures prevalent in the upper mantle, such euhedral diamond crystals are inferred to have formed at depths greater than 100 miles.

Because they are highly resistant to weathering and due to their relatively high specific gravity (3.5 g/cc) most diamonds have been recovered from stream gravels or beach deposits. However, South Africa, the largest producer of diamonds in the world, still produces many of its diamonds from kimberlite pipes. Other countries that are major producers of diamonds are Australia, Tanzania, Brazil, and the former Soviet Union. Diamonds from the Ozark Mountains in Arkansas are found in surface deposits ("yellow ground") of kimberlite pipes of Cretaceous age.

The diamond-bearing kimberlite is blasted out, transported, crushed, concentrated, and, in the final stages of concentration, the hydrophilic (resistance to wetting) properties of diamond are utilized. Because diamond resists wetting, it will stick to grease even after being exposed to water. The concentrate is wetted and passed over tables that have been greased so that the diamonds stick, whereas the kimberlite rock passes through and is eliminated.

The great hardness and high *refractive index* and *dispersion* make it the most prized and highly valued of gemstones. The high values of these two parameters result in a beam of white light splitting into its constituent colors, giving the diamond its "fire." The value of a cut diamond varies widely depending on factors such as clarity, color, lack of imperfections, and quality of faceting. Diamond powder and crystal fragments are used widely as abrasives. Synthetic diamonds are produced in large quantity as abrasive material; they are unsuitable for gemstones because of their small size.

Graphite is found in metamorphic rocks such as slates, schist, and gneiss, and in some hydrothermal deposits. In both instances, the carbon was probably derived from an organic source and subsequently converted to graphite by heat. The most common use of graphite is in the manufacture of pencil lead. Graphite is also widely used as a lubricant in both dry powdered form and mixed with oils. Because of its refractory properties, graphite is used in making crucibles (vessels having a high melting point in which molten steel is carried) for the steel industry.

3.3 SULFIDES AND SULFOSALTS

The sulfides and sulfosalts (Table 3.2) occur primarily in *hydrothermal ore deposits*—deposits formed by the crystallization of elements concentrated and carried in a hot water solution produced as magma cooled or by water flowing through hot fractured rock. These types of deposits occur in a large number of geologic environments and

TABLE 3.2
The Sulfide and Associated Minerals

Mineral Name	Chemical Composition	Hardness	Specific Gravity	Uses	Diagnostic Properties
Argentite	Ag_2S	2	7.0	Ore of Ag	Metallic black, sectile, high specific gravity.
Arsenopyrite	$FeAsS$	6	6.0	Ore of As	Silver metallic luster, black streak, prismatic crystals
Bornite	Cu_5FeS_4	3	5.0	Ore of Cu	Known as peacock ore for its purple and bluish colors on a tarnished surface.
Calaverite	$AuTe_2$	2.5	9.0	Ore of Au	Yellow to silver metallic luster, brittle
Chalcocite	Cu_2S	3	5.5	Ore of Cu	Black to lead–gray color, metallic luster, sectile.
Chalcopyrite	$CuFeS_2$	4	4.0	Minor Cu ore	Dark yellow color, metallic luster and greenish black streak.
Cinnabar	HgS	2.5	8.0	Ore of Hg	Red color. Usually forms earthy crusts on fracture surfaces.
Cobaltite	$(Co, Fe)AsS$	5.5	6.0	Ore of Co	Silver metallic luster, associated with other cobalt minerals.
Covellite	CuS	2	4.6	Ore of Cu	Dark blue, metallic luster and platy cleavage.
Galena	PbS	2.5	7.5	Ore of Pb	Cubic crystals and cleavage, metallic gray color, high specific gravity.
Marcasite	FeS_2	6	5.0	N/A	Pale yellow metallic luster tarnishing black. Cockscomb crystal aggregates.

TABLE 3.2
Continued

Mineral Name	Chemical Composition	Hardness	Specific Gravity	Uses	Diagnostic Properties
Millerite	NiS	3.5	5.5	Minor Ni ore	Light yellow acicular crystals with metallic luster. Forms in cavities.
Molybdenite	MoS_2	1	4.0-5.0	Ore of Mo	Bluish-gray metallic luster, soft, greasy feel.
Niccolite	NiAs	5	7.0-8.0	Ore of Ni	Brownish–red metallic luster, high specific gravity.
Orpiment	As_2S_3	2	3.5	N/A	Bright yellow color and association with Realgar.
Pentlandite	$(Fe, Ni)_9S_8$	4	5.0	Major ore of Ni	Light metallic bronze color, nonmagnetic.
Pyrite	FeS_2	6	5.0	Minor source of sulfur	Light golden yellow, metallic luster, striated cubes or pyritohedrons.
Pyrrhotite	$Fe_{1-x}S$	4	4.5	N/A	Dark bronze metallic luster, magnetic, black streak.
Realgar	AsS	2	3.5	N/A	Red to orange color, commonly earthy.
Sphalerite	ZnS	4	4.0	Ore of Zn	Yellowish brown, resinous luster; six directions of cleavage.
Stibnite	Sb_2S_3	2	4.5	Major ore of Sb	Radiating bladed crystals, lead gray, metallic luster.
Sylvanite	$(Au, Ag)Te_2$	2	8.0	Ore of Ag and Au	Metallic silver color, crystals commonly bladed.

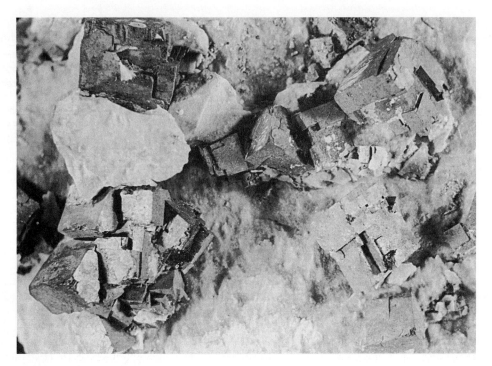

FIGURE 3.5
Cubic crystals of galena on limestone. Crystal sizes range from 1 to 2 cm in diameter.

display a wide range of chemical composition and rock type associations. The water-rich solutions had temperatures ranging from 50°C to 600°C. Some hydrothermal deposits have a limited chemical and mineralogical compositional range; for example, gold lode deposits are typically composed of gold, quartz, and pyrite. Other hydrothermal deposits are a rock-hounder's delight, containing a cornucopia of minerals, such as the lead-silver-zinc deposits common in the Cordillera (Rocky Mountains) of North America. Typical ore minerals in these deposits are galena (PbS; Figure 3.5) and sphalerite (ZnS), as well as accessory pyrite (FeS_2; Figure 3.6) and chalcopyrite ($CuFeS_2$), and a wide range of other ore and gangue minerals.

Physical Properties

As a group, the sulfides have a high specific gravity, ranging from four to eight g/cc, and a metallic luster. The exceptions to this rule are cinnabar (HgS), which is red and may have a vitreous luster (in crystals) or an earthy luster, and sphalerite (ZnS) which has a resinous luster. The sulfide deposits are the source of many metals. The sulfosalts are a much more diverse group with a wide variety of physical properties, although in general they all possess an above average specific gravity of 4.0 to 6.0 g/cc.

FIGURE 3.6
Crystals of pyrite showing the metallic luster and striated typical faces.

The tellurides are a small group of very important minerals because the major cations in this mineral group are gold or silver. The two most most common species, calaverite and sylvanite, are both characterized by metallic luster, yellow to silver-white color, high specific gravity (8.0 to 9.0 g/cc) and low hardness (1½ to 2½).

3.4 FORMATION OF A HYDROTHERMAL ORE DEPOSIT

Probably the most abundant type of ore deposit in the mountain ranges of western North and South America is the Ag-Pb-Zn deposit (Table 3.3). The association of these three metals recurs in deposits throughout the region. Deposits from Mexico, Idaho, and Colorado are described in the following paragraphs.

The Santa Eulalia District, Chihuahua, Mexico

The Santa Eulalia District, located in a series of low limestone hills just a few miles southeast of the city of Chihuahua was discovered in the late 16th century. It is one of the richest silver districts in Mexico. The ore deposits occur as massive replacements of Cretaceous Period (Appendix I) limestone along fracture zones and fissures.

TABLE 3.3
Typical Ore and Gangue Minerals in a Cordilleran Pb-Zn-Ag Hydrothermal Deposit

Mineral	Composition
Ore Minerals in the Sulfide Zone	
Argentite	Ag_2S
Bornite	Cu_5FeS_4
Chalcopyrite	$CuFeS_2$
Cinnabar	HgS
Covellite	CuS
Galena	PbS
Native Silver	Ag
Sphalerite	ZnS
Tetrahedrite	$Cu_{12}Sb_4S_{13}$
Ore Minerals in the Oxidized Zone	
Anglesite	$PbSO_4$
Azurite	$Cu_3(Co_3)_2(OH)_2$
Cerussite	$PbCO_3$
Chrysocolla	$CuAl_2H_2Si_2O_5(OH)_4$
Cuprite	Cu_2O
Malachite	$Cu_2CO_3(OH)_2$
Smithsonite	$ZnCO_3$
Accessory Minerals	
Barite	$BaSO_4$
Calcite	$CaCO_3$
Dolomite	$CaMg(CO_3)_2$
Fluorite	CaF_2
Gypsum	$CaSO_4 \cdot 2H_2O$
Opal	$SiO_2 \cdot nH_2O$
Pyrite	FeS_2
Pyromorphite	$Pb_2(PO_4)_3Cl$
Quartz	SiO_2
Wulfenite	$PbMoO_4$
Hematite	Fe_2O_3

These near-vertical fracture zones provided easy access for post-Cretaceous hydrothermal solutions rising through the limestone sequence. At favorable horizons, large areas of limestone have been completely replaced by hydrothermal minerals forming *mantos*, flat-lying bedded ore deposits. The vertical ore zones are called *chimneys* because of their geometry. Oxidized ores containing oxides, hydroxides, and carbonates of the hydrothermal metals are abundant in the deposit's upper levels. At lower levels, moderate temperature sulfide ores (galena, sphalerite, pyrite) and higher temperature silicate mineralization (actinolite, magnetite, and fayalite with minor base metal sulfides) are present. The bulk of the mining operation is concentrated in the sulfide ores. The ore grade averages nine percent to 10 percent each of lead and zinc and 10 to 12 ounces of silver per ton of ore. The silver is contained in galena, where the silver ion, of similar size and charge to the Pb ion, substitutes in galena's atomic structure. In the richest zones, six kilograms of silver are produced per ton of galena ore.

Numerous dikes, sills, and irregular intrusions of igneous rock in the vicinity of Santa Eulalia are evidence of intrusive activity in the area since the end of the Cretaceous Period. These intrusions are the probable source of heat for the hydrothermal solutions. The water was probably derived in part from the water contained in the magma, and in part from groundwater. The metals were also derived from a variety of sources.

The Coeur d'Alene District, Idaho

The Coeur d'Alene District has produced lead, zinc, and silver as well as copper and antimony since 1885. The ore minerals of this district include sphalerite, galena, bornite, chalcocite, stibnite, and tetrahedrite. The ore is distributed in zones that are concentrated along fractures and shears in a sedimentary sequence of Precambrian age known as the Belt Series. The mineralization is adjacent to igneous intrusions, and is zoned, having mainly gold mineralization in the north, lead and zinc in the central part, and copper and silver in the south. The Sunshine Mine, the largest mine in the district, has Pb-Zn ore averaging 10 percent combined Pb-Zn and some silver occurring in tetrahedrite. Annual production from this mine averages approximately 5 million ounces of silver from ore containing 28 to 48 oz./ton.

The Cripple Creek District, Colorado

A cowboy, Bob Womack, discovered gold in the Cripple Creek area in 1890, but sold his claim before the full extent of his discovery became known. The gold mineralization is chiefly in the form of tellurides with associated pyrite, fluorite, and quartz. All of the mines are within or adjacent to the so-called Cripple Creek Basin, a volcanic caldera of irregular shape that is 4 miles long and 2 miles wide. That this steep-walled basin formed in Miocene (Appendix I) time has been determined by the age of the sediments and fragmental volcanic rocks (breccias) filling it. The nature of the rock in the basin reveals how important to the formation of ore deposits are the events that occur prior to mineralizing fluid invasion. The basin was first filled by fragmental rocks that were then cut by a series of faults and shear zones. These shear zones extend into the underlying basement rock, allowing the upward movement of mineralizing fluids. The sediments in the basin were a veritable sponge allowing the precipitation of hydrothermal minerals. The result was a district that has yielded more than 21 million troy ounces of gold since 1891.

These mining districts all had a source of heat, usually a nearby igneous body, and experienced extensive fracturing of the rock surrounding the heat source. These two conditions are probably the most essential factors for forming an economic hydrothermal ore deposit.

3.5 THE CONCEPT OF MINERAL ASSOCIATIONS

Study of a large number of ore deposits has revealed that some minerals typically occur in association with certain others. One of the most common of such associations is Pb-Zn-Ag mineralization. These minerals occur in association with a large number of others in noneconomic concentrations (gangue minerals) such as fluorite, quartz, calcite, wulfenite, and pyrite. Another common mineral suite is the skarn assemblage (Chapter 11), forming adjacent to igneous intrusions in carbonate rocks. Skarn mineral assemblages consist of calcium and magnesium silicates. Another example is the association of cobalt and nickel. A corollary to mineral association concept is that the major ore mineral chemistry determines the occurrence of the gangue minerals. For example, the presence of sphalerite, a zinc sulfide, is an indicator of other zinc minerals, such as smithsonite, and of other sulfide minerals. Thus, it is important to keep in mind some of the more common mineral associations as well as a general idea of the chemical composition of each mineral.

4
The Industrial Minerals

This group of minerals includes more species than any other group except for the silicate minerals. The minerals within the group are classified by the anion that, with few exceptions, is oxygen or a combination of oxygen and another element. Included in this group are the oxides ($O^=$), hydroxides (OH^-), carbonates (CO^3), sulfates (SO^4), borates (BO^3), halides (F^- or Cl^-), tungstates (WO^4), molybdates (MoO^4), phosphates (PO^4), nitrates (NO^3), arsenates (AsO_4) and vanadates (VO^4). This diverse group includes ore minerals such as hematite and magnetite; gemstones, such as rubies and sapphires (the mineral corundum), and alexandrite (the mineral chrysoberyl); and industrial minerals used in the manufacture of glass (the borates), fertilizer (the phosphate and potash minerals), and building materials (calcite and gypsum).

4.1 THE OXIDES AND HYDROXIDES

As a group, the *oxide minerals* are relatively hard, have a moderate to high specific gravity (Table 4.1), and generally occur in small amounts in igneous and metamorphic rocks as accessory minerals. Because of their resistance to weathering, some may form placer deposits. *Hydroxides* are softer and have a lower specific gravity than the oxides and form by alteration or chemical weathering of other minerals.

TABLE 4.1
The Oxide and Hydroxide Minerals

Mineral Name	Chemical Composition	Hardness	Specific Gravity	Uses	Diagnostic Properties
Boehmite	AlO·OH	3	2.5	A major constituent of bauxite	As pisolitic aggregates in bauxite —the major ore of Al.
Brucite	$Mg(OH)_2$	2.5	2.4	A minor source of Mg	Light gray, vitreous to pearly luster.
Cassiterite	SnO_2	7	7.0	Ore of Sn	High specific gravity, adamantine luster. Also as wood tin.
Chromite	$FeCr_2O_4$	5	4.5	Ore of Cr	Black submetallic luster, granular, dark brown streak.
Columbite	$(Fe, Mn)Nb_2O_6$	6	5.0-8.0	Source of Nb. Forms solid–soln series with Ta	Orth. crystal form, high specific gravity dark brown to black, submetallic luster.
Corundum	Al_2O_3	9	4.0	Gemstones, abrasives	Hexagonal barrel, shaped crystals, superior hardness
Cuprite	Cu_2O	4	6.0	Minor Cu ore	Red metallic luster, isometric crystal form
Diaspore	AlO·OH	6	3.0	Major constituent of bauxite	Bladed light gray, vitreous crystals. Pisolitic in bauxite.
Goethite	FeO·OH	5	4.0	Iron ore	Yellow to brown, earthy to adamantine luster, usually in radiating fibrous aggregates.
Hematite	Fe_2O_3	2-6	2.5-5.0	Major ore of Fe	Reddish brown (earthy variety) to metallic black (specular variety) Hexagonal tabular crystals.

TABLE 4.1
Continued

Mineral Name	Chemical Composition	Hardness	Specific Gravity	Uses	Diagnostic Properties
Ilmenite	$FeTiO_3$	6	4-5	Ti ore, paint pigment	Black metallic luster. May be slightly magnetic.
Magnetite	Fe_3O_4	6	5.0	Ore of Fe	Black metallic luster. Strongly magnetic, forms octahedrons.
Manganite	$MnO(OH)$	4	4.5	Minor Mn ore	Metallic black luster, dark brown streak.
Psilomelane	$(Ba, Mn)_3(O, OH)_6Mn_8O_{16}$	6	4.0	Ore of Mn	Black submetallic luster. Forms botryoidal masses.
Pyrolusite	MnO_2	2	4.5	Ore of Mn	Black metallic to submetallic luster, in thin radiating crystals and as dendrites.
Rutile	TiO_2	6	4.0	Source of Ti and TiO_2	Reddish-brown adamantine luster, prismatic or acicular crystals.
Spinel	$MgAl_2O_4$	8	4.0	Genstone	Vitreous octahedral crystals, with superior hardness, various colors.
Uraninite	UO_2	5.5	8.0	Major U ore	Black submetallic luster, radioactive.
Zincite	ZnO	4	6.0	Source of Zn	Deep red submetallic luster. Associated with Franklinite at Franklin, N. J.

FIGURE 4.1
Crystal of corundum showing the typical hexagonal prism; 1.5 cm in diameter.

Corundum

Red rubies and blue sapphires are both varieties of the mineral *corundum* (Al_2O_3). Trace amounts of Cr and Ti produce the red color of rubies and the blue color of sapphires. The hardness of corundum (H = 9) is second only to that of diamond, making corundum a valuable abrasive material and gem mineral. The high specific gravity (4.0 g/cc) causes this mineral to be concentrated in placer deposits. Corundum crystallizes in the *hexagonal system* and typically forms plates or barrel-shaped crystals with six sides (Figure 4.1). In some occurrences, microscopic crystal impurities of rutile crystallize in the corundum in a hexagonal pattern at 60 degree angles to each other. The light reflected from such crystals forms a six-pointed star, and thus, corundum is said to display *asterism*, a property seen in star sapphires and rubies.

Historically the highest quality rubies and sapphires have come from Burma, but several other localities in southern and southeast Asia, including Thailand, Indonesia, Cambodia, as well as Kashmir in the Himalayas, have also produced these gems. More recently, a large number of gemstones have been found in Afghanistan, where mining has been facilitated by the Russian bomb craters resulting from the war in Afghanistan. In the United States several varieties of corundum have been found in the Appalachian and Rocky Mountains.

Emery is a fine-grained abrasive mixture of magnetite and corundum. It is a valuable industrial abrasive, although much of it is synthetically produced today for abrasive disks, wheels, and emery boards.

Hematite and Magnetite

Hematite and *magnetite* are the most important ores of iron, and the principal minerals on which the industrial revolution was based. In the present age of environmental concern, iron and steel are some of the most easily and frequently recycled materials. Hematite (Fe_2O_3) crystallizes in the hexagonal system and occurs in two distinct forms: a soft *earthy* variety, typically red or brown and with variable hardness (due to mixtures with other minerals); and, a harder (H = 6) *specular* variety that is black, has a metallic luster, and commonly displays a hexagonal structure in its platy crystals. Both varieties, however, have a characteristic brick red streak.

Ancient humans used and valued hematite, not as a source of iron, but as a bright red pigment used as body paint and as paint for the walls of Mayan and Egyptian tombs and cave paintings as much as 30,000 years old. The specular variety of hematite is today sometimes polished and set in a silver setting.

Magnetite (Fe_3O_4) is an *isometric* mineral, typically forming octahedrons. Its most outstanding characteristic is its strong magnetism. Natural magnets of magnetite are known as *lodestones*. Like hematite, it is an important iron ore and occurs in a variety of geologic environments. For example, it is a widespread *accessory* mineral in igneous and metamorphic rocks, meaning it occurs in minor amounts (one to five percent by volume). It is also a common placer mineral in some localities and may be a major constituent of unusual black river and beach sands. Magnetite also occurs in contact metamorphic rocks, that group of rocks metamorphosed by their contact with igneous intrusions of magma. The highest-grade deposits of magnetite ore are produced by magmatic segregation, a process in which a magma's iron-rich portion segregates from its silica-rich fraction and is injected or extruded into the surrounding rock. Examples of such deposits occur in Kiruna, Sweden, and Cerro Mercado, Durango, Mexico.

Ilmenite and Rutile

Ilmenite ($FeTiO_3$; hexagonal) and *rutile* (TiO_2; tetragonal) are the major sources of titanium metal. Much of the rutile mined is used in the form of TiO_2 as paint and ceramic pigment. Both minerals have a hardness of about six and a specific gravity of 4.0 to 4.5 g/cc: ilmenite is metallic black, much like magnetite, and rutile has a reddish-brown adamantine luster. Ilmenite, unlike magnetite, is weakly magnetic.

FIGURE 4.2
Pyrolusite dendrites on fracture surface. This inorganic growth form is commonly misinterpreted as a fossil. Photograph width, 6 cm.

Ilmenite is a common accessory mineral in igneous rocks; rutile, less so. Both minerals are major constituents of black sands, occuring in association with magnetite, zircon, and monazite.

Cassiterite

Cassiterite (SnO_2; *tetragonal*) is the principal ore of the important alloy metal, tin. It possesses a brilliant adamantine luster, an unusually high specific gravity of about seven, and is almost as hard as quartz ($H = 7$). Its primary occurrence is in hydrothermal veins formed at high temperature (500 to 600°C), and usually associated with granites. However, much of the world's tin production comes from alluvial (stream deposited) rocks in the form known as *stream tin*.

Tin was one of the first metals utilized by humans, being an essential constituent of the alloy metal bronze. The Phoenicians regularly shipped tin from deposits at Cornwall, England and possibly Turkey, between 1,500 and 2,000 B.C.. A bronze rod containing nine percent tin found in Egypt dates back to 3,700 B.C. (Jensen & Bateman, 1981). Today the major world producers of tin are Malaysia and

Bolivia. No large deposits of tin ore occur in the United States, despite its consumption of nearly one-third of the world's production.

Other Oxide Minerals

Other *oxides* include *uraninite* (UO_2), a major ore of uranium; *chromite* ($FeCr_2O_4$), the major ore of the element *chromium* found associated with ultramafic rocks in the lower portion of stratiform complexes; and *chrysoberyl* ($BeAl_2O_4$), a relatively rare mineral that includes the gem variety, alexandrite. *Pyrolusite* (MnO_2) is the most common of the manganese ore minerals. Characteristically, it has a metallic luster, black color, is soft, and has an acicular habit. Pyrolusite can also be found as dendritic growths on fracture surfaces (Figure 4.2).

The Hydroxide Minerals

Within the *hydroxide group* are the minerals brucite, psilomelane, goethite, and bauxite. *Brucite* is a white to light green, soft (H = 2) mineral, having a lower than average specific gravity. Usually associated with serpentine, it is used in refractory materials.

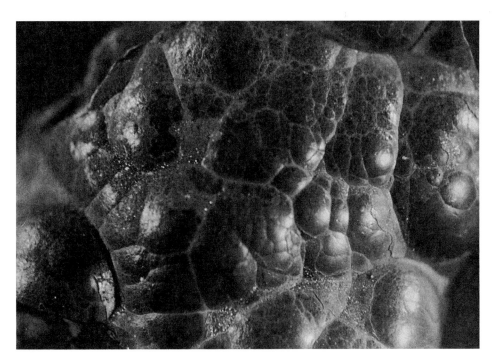

FIGURE 4.3
Botryoidal texture on goethite mass.

Psilomelane ($BaMn_9O_{16}(OH)_4$) is a hard (H = 5 to 6) mineral with a black submetallic luster. It is an important ore of manganese, a vital component of most types of steel.

Goethite (FeOOH) is an iron hydroxide usually found in radiating fibrous aggregates forming botryoidal masses (Figure 4.3). It resembles hematite and is also an ore of iron. The mineral was named in honor of Goethe, the German poet.

Bauxite is a rock consisting of aluminum oxides and hydroxides, principally gibbsite, boehmite, and diaspore. Bauxite is important because it is the ore of aluminum. It is a weathering product of feldspars and clays formed in tropical environments characterized by abundant rainfall and intense leaching of the soil horizon.

TABLE 4.2
The Carbonate Minerals

Mineral Name	Chemical Composition	Hardness	Specific Gravity	Uses	Diagnostic Properties
Hexagonal Group					
Calcite	$CaCO_3$	3	3.0	Source of lime, manufacture of cement	Clear to white, vitreous luster, rhombohedral cleavage, effervesces in dilute HCl.
Dolomite	$CaMg(CO_3)_2$	4	3.0	Building stone	Light pink rhomb. crystals with curved faces. Powdered mineral effervesces in HCl.
Magnesite	$MgCO_3$	3-5	3.0	Sometime ore of Mg	Associated with other Mg minerals.
Rhodocrosite	$MnCO_3$	4	4.0	Minor Mn ore	Reddish pink to brown, vitreous crystals. High specific gravity.
Siderite	$FeCO_3$	3.5	4.0	Iron Ore	Brown vitreous rhombs, usually with curved faces.
Smithsonite	$ZnCO_3$	4	4.5	Ore of Zn	Vitreous luster. Color variable, high specific gravity.

Although other minerals, such as feldspars and clays, contain a large percentage of aluminum, bauxite is the most amenable to Al extraction. Another important use of bauxite is in the production of synthetic alumina (Al_2O_3).

4.2 THE CARBONATES

The *carbonate* minerals can be subdivided into three groups based on their crystallography (Table 4.2); the *rhombohedral, orthorhombic,* and *monoclinic* groups. The chemical composition of the rhombohedral carbonates can be represented on the triangular diagram shown in Figure 4.4. As a group, the rhombohedral carbonates

TABLE 4.2
Continued

Mineral Name	Chemical Composition	Hardness	Specific Gravity	Uses	Diagnostic Properties
Orthorhombic Group					
Aragonite	$CaCO_3$	4	3.0	As for calcite. Source of lime, manufacture of cement	White, vitreous luster, slightly harder and denser than calcite.
Cerussite	$PbCO_3$	3.5	6.5	Pb ore	White or gray, vitreous luster also adamantine. Very high specific gravity.
Strontianite	$SrCO_3$	4	3.6	Sr ore	Radiating, acicular, vitreous luster, white to gray crystals.
Witherite	$BaCO_3$	3.5	4.5	Minor Ba ore	White to gray, vitreous luster.
Monoclinic Group					
Azurite	$Cu_3(CO_3)_2(OH)_2$	3.5	3.8	Minor Cu ore	Dark blue, vitreous luster, effervesces in HCl.
Malachite	$Cu_2CO_3(OH)_2$	3.5	4.0	Ornamental material, Cu ore	Dark green, vitreous luster, typically forms botryoidal masses, fibrous.

FIGURE 4.4
Triangular composition diagram showing the composition of the rhombohedral Ca-Fe-Mg carbonates.
Source: Klein C. and C. S. Hurlbut (1985). *Manual of Mineralogy*. 20th ed. New York: John Wiley & Sons. Fig. 10.5, p. 330.

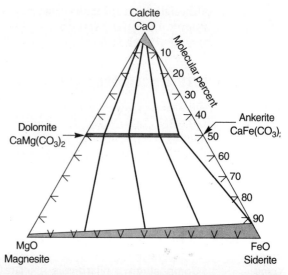

FIGURE 4.5
Fragment of optical calcite, so-called because of its water-clear appearance, showing the rhombohedral crystal form. Crystal is 5 cm in diameter.

FIGURE 4.6
Small (3–6 mm diameter) light-colored crystals of dolomite in a cavity.

exhibit *rhombohedral* cleavage—in three directions, not at right angles—and most effervesce (bubble CO_2) in cold dilute hydrochloric acid (HCl). Magnesite and rhodocrosite effervesce in hot HCl.

Calcite ($CaCO_3$) occurs in a large number of crystal forms of the *trigonal system*, a subset of the hexagonal system, and exhibits rhombohedral cleavage (Figure 4.5). Calcite is widely used in the production of lime and cement and also widely used for its chemical properties. Magnesite ($MgCO_3$) has been mined as an ore for Mg metal, but most is now extracted from brines and seawater. Mg is the lightest known metal (specific gravity = 1.7) and is used for light, yet strong, alloys. *Siderite* ($FeCO_3$) is a minor ore of iron recognized by its brown color and rhombohedral cleavage. *Rhodocrosite* ($MnCO_3$) is a minor ore of manganese and is a sought-after collector's mineral as crystals and stalagmites with a distinctive pink color. *Smithsonite* ($ZnCO_3$) is a minor ore of zinc and is another collector's prize especially where colored by minor impurities forming varieties such as the Kelly green smithsonite from the Magdalena District of central New Mexico. *Dolomite* ($CaMg(CO_3)_2$), abundant in the sedimentary rock *dolostone*, is a widely used building stone. It is also found as well-formed crystals within the cavities of hydrothermal ore deposits (Figure 4.6).

The orthorhombic calcium carbonate *aragonite* is the common $CaCO_3$ of seashells and pearls, and is also formed in hot springs and in cave environments. It

FIGURE 4.7
White, tabular crystals of barite. Lincoln County, New Mexico.

is slightly harder and has a higher specific gravity than calcite. It is a metastable mineral, however, and transforms to calcite after about 60 million years. Witherite ($BaCO_3$), strontianite ($SrCO_3$), and cerussite ($PbCO_3$) are typically formed in hydrothermal veins or as low-temperature alteration products of various minerals within them. Minerals with these are barite, celestite and galena.

The monoclinic carbonates, *malachite* and *azurite*, are a colorful pair of hydrous copper carbonates usually found in the upper oxidized portion of copper ore deposits. The deep green color of malachite in particular has been widely used as an ornamental material, and both have been utilized as ores of copper and pigments.

4.3 THE SULFATES

The *sulfates* are a small mineral group containing some very common, widely used minerals (Table 4.3). Barite ($BaSO_4$) is identified by its tabular crystals and its high specific gravity (Figure 4.7). Barite is a common accessory or gangue mineral in many hydrothermal ore deposits. It is a major source of Ba for chemical uses and is widely

FIGURE 4.8
A vein of satin-spar gypsum showing the fibrous habit of this variety of gypsum. Vein is 5 cm thick.

and extensively used in its mineral form as a component of drilling muds used in the petroleum industry. Barite's high specific gravity increases the specific gravity or density of the drilling mud, and aids in the prevention of blowouts.

Celestite ($SrSO_4$) is one of the least common of the sulfate minerals. It is distinguished by its light blue, celeste color, and higher than average specific gravity. It is a source of strontium. *Anhydrite* ($CaSO_4$) is the anhydrous (water-free) form of calcium sulfate. It is usually massive and is distinguished from the hydrous form, gypsum ($CaSO_4 \cdot 2H_2O$), by its greater hardness. Anhydrite is a minor source of sulfur. *Gypsum* is much more abundant than anhydrite in the near surface environment. It is characterized by low hardness (H = 2) and low specific gravity (2.2 g/cc) and occurs in three varieties. *Alabaster* gypsum is a fine-grained, massive variety, and is usually white. *Satin spar* gypsum is a fibrous variety with a silky or pearly luster (Figure 4.8). *Selenite* gypsum is a transparent crystalline variety exhibiting perfect cleavage in one direction. It is sometimes twinned to form swallowtail or fishtail crystals (Figure 4.9). Gypsum forms by evaporation from brines. It is used in many products, mainly the manufacture of wallboard.

4.4 THE BORATES

Of the more than 100 described borate minerals only four or five can be considered common. The building block for the borates are BO_3 triangles linked together with Na or Ca ions. As a group, the common borates are white or light-colored, soft (H = 2 to 4), have specific gravities less than three, and show prismatic or fibrous habit (Figures 4.10, 4.11, and Table 4.3).

TABLE 4.3
The Sulfate, Borate, and Halide Minerals

Mineral Name	Chemical Composition	Hardness	Specific Gravity	Uses	Diagnostic Properties
Sulfate Minerals					
Anglesite	$PbSO_4$	3	6.0	Minor Pb ore	White to gray adamantine luster, associated with galena.
Anhydrite	$CaSO_4$	3.5	3.0	Possible S source	Generally as granular to white masses, assoc. with gypsum.
Gypsum	$CaSO_4 \cdot 2H_2O$	2	2.2	Wall board, plaster of Paris	3 varieties of gypsum are recognized: selenite–transparent satin spar–fibrous alabaster–fine-grained
Barite	$BaSO_4$	3.5	4.5	Major Ba ore. Used in drilling muds	Tabular, vitreous, white to colorless crystals, high specific gravity.
Celestite	$SrSO_4$	3.5	4.0	Source of Sr	Tabular, vitreous to pearly bluish crystals. High specific gravity.
Borate Minerals					
Borax	$Na_2B_4O_5(OH)_4 \cdot 8H_2O$	2.5	2.0	Medicines, flux. Source of B	Prismatic, vitreous, white to colorless crystals. Low hardness and specific gravity.

The Industrial Minerals

Borate minerals form principally in bedded deposits beneath old playa (hypersaline) lake beds in arid or semi-arid environments, in playa lake brines, and around volcanic vents and hot springs. The largest and best known deposits are in the Death Valley area of southeastern California, and in Turkey and Argentina. The borate minerals are refined into the most useful form, borax, which is used in manufacturing glass, oils, and varnishes, and gives leather a smooth, soft finish.

TABLE 4.3
Continued

Mineral Name	Chemical Composition	Hardness	Specific Gravity	Uses	Diagnostic Properties
Colemanite	$CaB_3O_4(OH)_3 H_2O$	4	2.5	Source of borax	Hardest of the common borate minerals. White, vitreous prismatic crystals.
Kernite	$Na_2B_4O_6(OH)_2 \cdot 3H_2O$	3	2.0	Source of borax	Perfect cleavage in 2 directions forms white vitreous to chalky splinters.
Ulexite	$NaCaB_5O_6(OH)_6 \cdot 5H_2O$	1-2.5	2.0	Source of borax	Fibrous or acicular, white pearly luster.
Halide Minerals					
Cerargyrite	$AgCl$	2.5	5.5	Ore of Ag	Gray to colorless, wax-like masses, sectile, high specific gravity.
Fluorite	CaF_2	4	3.2	Flux in steel making, HF acid	Usually as cubes, cleaves as octahedrons. Color variable, commonly purple; vitreous.
Halite	$NaCl$	2.5	2.1	Industrial chemical	Cubic crystals and cleavage, salty taste.
Sylvite	KCl	2	2.0	Source of K	White, vitreous crystals usually in shades of red. Bitter taste.

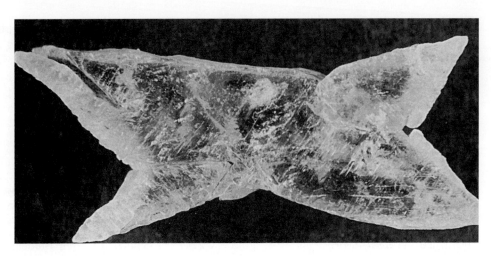

FIGURE 4.9
Fishtail twins of selenite gypsum, 10 cm length.

FIGURE 4.10
Fibrous ulexite, a common borate, southeastern California; fibers 2 cm long.

FIGURE 4.11
Radiating acicular crystals of colemanite, a common borate. Sprays of crystals are 5 cm in diameter.

4.5 THE HALIDE MINERALS

The *halides* are a small group containing three widely used industrial minerals: halite, sylvite, and fluorite. *Halite* (NaCl) and *sylvite* (KCl) may occur together in bedded marine deposits formed in sedimentary basins where restricted circulation of seawater promoted evaporation and thus concentrated brines from which evaporite minerals formed. They can be distinguished by the more bitter taste of sylvite. Halite (Figure 4.12) is used in the chemical industry as a source of sodium compounds and, in the mineral form, in various ways ranging from table seasoning, to the melting of ice on roads and highways, to industrial processes. Sylvite is the major source of potassium, an essential ingredient in fertilizers. More recently it has come to be used in seasonings as salt substitutes for persons who must restrict their intake of sodium (salt).

Halite and sylvite deposits occur in the Permian Basin of west Texas and southeast New Mexico, in the Paradox Basin of southeastern Utah, in the Michigan Basin near Detroit, and in the Williston Basin of Saskatchewan, Canada. Other potassium salts which occur in these deposits in association with sylvite are carnallite (KMgCl$_3$ • 6H$_2$O), polyhalite (K$_2$Ca$_2$Mg(SO$_4$)$_4$ • 2H$_2$O) and kainite (KMg$_2$(Cl(SO$_4$)Cl • 3H$_2$O).

Fluorite is a very common gangue mineral in hydrothermal ore deposits. It is found in quantity in the Mississippi Valley deposits in the mid-continent region of the United States, and in many small deposits in western North America. Its color is highly variable, ranging from green to blue, purple, and colorless. Typically it forms cubes and the combination of cubic crystal form, octahedral cleavage, and higher than average specific gravity make this mineral readily identifiable.

Fluorite (CaF$_2$) is used mainly as a flux in manufacturing steel; 14 pounds of fluorspar (fluorite's industrial name) are necessary to produce each ton of steel. It is also used in manufacturing hydrofluoric acid, a strong acid used in etching glass, and fine optics for microscopes.

Other halide minerals include the mineral *cryolite* (Na$_3$AlF$_6$) distinguished by its singular significant ore deposit: Ivigtut on the west coast of Greenland. Formerly used in producing aluminum from bauxite, it has now been replaced by a synthetic substitute.

4.6 THE TUNGSTATES, MOLYBDATES, PHOSPHATES, AND VANADATES

The only two *tungstates* considered common are *scheelite* and *wolframite* (Table 4.4). Scheelite (CaWO$_4$), the more common tungstate in the United States, occurs in granitic pegmatites (extremely coarse-grained, fluid-rich plutonic bodies) and in contact metamorphic deposits. Its most distinctive physical property is its purplish-white fluorescence in short-wave ultraviolet light. Wolframite ((Fe,Mn)WO$_4$) is dis-

FIGURE 4.12
Cubic halite crystal, 4 cm edge.

TABLE 4.4
The Tungstate, Molybdate, Phosphate, and Vanadate Minerals

Mineral Name	Chemical Composition	Hardness	Specific Gravity	Uses	Diagnostic Properties
Apatite	$Ca_5(PO_4)_3(F,Cl)$	5	3.2	Source of PO_4	Hexagonal, prismatic, usually light green, vitreous luster.
Autunite	$CaUO_2(PO_4)_2 \cdot 10\text{-}12\,H_2O$	2	3.0	Ore of U	Tabular, yellow to green, vitreous crystals in scaly aggregate, radioactive.
Carnotite	$K_2(UO_2)_2(VO_4)_2 \cdot 3H_2O$	1-2	5.0	Ore of U and V	Bright yellow mineral, usually as powdery crusts, radioactive.
Monazite	$(Ce,La,Y,Th)PO_4$	5.5	5.0	Source of ThO_2	Resinous to adamantine luster, yellow to brown color, radioactive.
Scheelite	$CaWO_4$	5	6.0	Minor W ore	Vitreous to adamantine, white to brown crystals. Fluoresces purple-white in UV radiation.
Turquoise	$CuAl_6(PO_4)_4(OH)_8 \cdot 4H_2O$	6	2.7	Gemstone	Blue to bluish-green, waxy luster, massive.
Wavellite	$Al_3(PO_4)_2(OH)_3 \cdot 5H_2O$	3.5	2.5		Radiating, sperulitic aggregregates of yellowish green acicular crystals.
Wolframite	$(Fe,Mn)WO_4$	4.5	7.5	Main W ore	Bladed or tabular, black, submetallic crystals. High specific gravity.
Wulfenite	$PbMoO_4$	3	7.0	Minor Mo ore	Tabular, vitreous orange to yellow crystals. High specific gravity.

FIGURE 4.13
Tabular, dark orange crystals of wulfenite.

tinguished by its very high specific gravity, resinous luster, and bladed habit. It is found in pegmatites and high-temperature quartz veins. Most of the world's tungsten resources, occurring principally as wolframite, are located in China.

Wulfenite ($PbMoO_4$) is a colorful yellow to orange mineral of the tetragonal system, usually having tabular crystals (Figure 4.13). It is found in the upper oxidized portions of lead-bearing veins associated with other colorful minerals such as vanadinite and pyromorphite.

Apatite ($Ca_5(PO_4)_3(F,Cl)$) is the most common phosphate mineral, occurring as an accessory mineral in many rock types—in pegmatites, alkalic (aluminum-potassium-rich) rocks—and is the major mineral component of bone and teeth. Its most common colors in hexagonal crystalline form are green or brown (Figure 4.14). It is

FIGURE 4.14
Hexagonal crystals of apatite (7 cm in diameter), showing typical symmetry.

distinguished from other minerals with hexagonal crystals (quartz, beryl, tourmaline) by a hardness of five.

Monazite ((Ce,La,Y,Th)PO$_4$) is the major source of thorium oxide and an important source of rare earth elements (REE's; e.g., Nb,Y+). It is a common accessory mineral in granites, pegmatites, and gneisses, and commonly mined from their weathering products, which become major constituents of placer sands also containing rutile, zircon, and ilmenite.

The most widely prized phosphate mineral is *turquoise*. It is pearly blue or bluish-green and rarely forms crystals. Turquoise occurs in the upper parts of copper-bearing veins, originating by alteration of primary minerals. Some of the jewelry passed-off as turquoise is actually the copper silicate, chrysocolla, or artificial substitutes that are softer than turquoise and do not have its pearly luster. The Navajo and Hopi Nations of the southwestern United States have acquired a singular reputation for crafting superior quality turquoise and silver jewelry.

The phosphate mineral *wavellite* (Figure 4.15) forms distinctive yellow to green radiating aggregates of acicular crystals. It is a low-grade metamorphic mineral.

FIGURE 4.15
Radiating, yellowish-green crystals of wavellite. Crystal sprays are 2 cm in diameter.

The uranium-bearing minerals *carnotite* (a vanadate) and *autunite* (a phosphate; Table 4.4) are both of secondary origin, found in oxidized portions of uranium deposits. Both are typically a bright yellow to yellowish-green and soft. Autunite is more likely to form crystals, while carnotite typically forms a powdery aggregate on the bedding planes of sedimentary rocks and in fracture surfaces. Both are considered ores of uranium. Carnotite is found and mined throughout the Colorado Plateau Province of the southwestern United States.

5
The Rock-Forming Minerals:
The Silicates

Of every 100 atoms in Earth's crust, an average of 62 are oxygen and 21 are silicon. With this information alone, one would predict that the silicate minerals are the most abundant in the crust. Almost all of the minerals forming the igneous and metamorphic rocks of the continental crust are silicates, one of the most abundant minerals being SiO_2 (quartz). One should also note that, with the exception of Fe and Al, the common metallic elements upon which industrialized societies depend form a minor fraction of Earth's crust, thus making minerals such valuable, costly resources.

5.1 CRYSTAL STRUCTURE

The basic building block of the silicate minerals is the relatively small silicon ion surrounded by larger oxygen ions in a structure known as the SiO_4 tetrahedron. The silicon ion is located at the center, surrounded by the four closest oxygen ions at the tetrahedron's corners. Because there are more negative than positive charges in the SiO_4 tetrahedron, each oxygen ion has the potential to bond with other ions of one tetrahedron or to be shared among more than one tetrahedron.

The several structures in which tetrahedra can be configured (Figure 5.1), are, in order of increasing complexity: isolated tetrahedra, double tetrahedra, tetrahedra rings, chains, layers, and the most complex configuration of all, three-dimensional tetrahedral structures.

The *nesosilicates* or isolated tetrahedra (Figure 5.1a) include the *garnet* and the *olivine* groups. The silicon to oxygen ratio in the mineral formula is 1:4, or SiO_4.

The *sorosilicates* or double tetrahedra (Figure 5.1b) include the epidote group and the mineral vesuvianite. Because one of the oxygens is shared between the two tetrahedra, the silicon to oxygen ratio is 2:7 or Si_2O_7.

The *cyclosilicates* or tetrahedral rings form in three different ways: rings of three tetrahedra (Si_3O_9), four tetrahedra (Si_4O_{12}) or six tetrahedra (Si_6O_{18}) (Figure 5.1c). Benitoite ($BaTiSi_3O_9$) is a relatively rare mineral composed of rings of three tetrahedra. Axinite is an example of a mineral with a four-tetrahedral ring structure, and beryl and tourmaline are examples of cyclosilicates with rings of six tetrahedra.

FIGURE 5.1
Structure of the Silicates:
a) Nesosilicates and Sorosilicates; b) Inosilicates; c) Cyclosilicates; and d) Phyllosilicates.
From William H. Blackburn and William H. Dennen, *Principles of Mineralogy,* copyright © 1988 Wm. C. Brown Communications, Inc., Dubuque, Iowa. All Rights Reserved. Reprinted by permission.

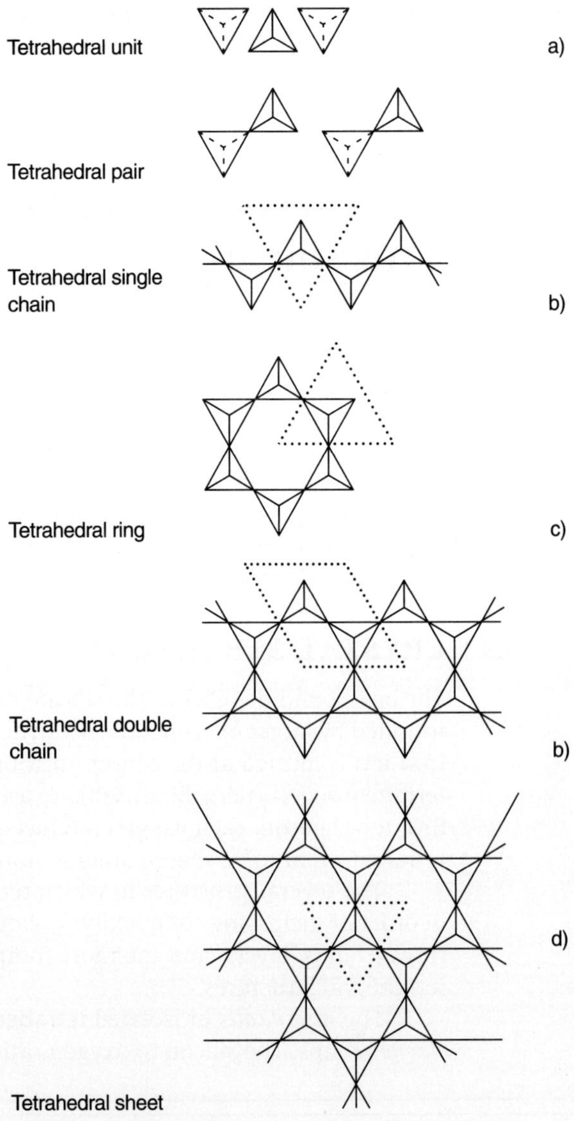

Tetrahedral unit — a)

Tetrahedral pair

Tetrahedral single chain — b)

Tetrahedral ring — c)

Tetrahedral double chain — b)

Tetrahedral sheet — d)

The *inosilicates*, or chains of tetrahedra, can be formed in two different ways: single and double chains (Figure 5.1d). The *pyroxene group* has a structure comprised of single chains with a silicon to oxygen ratio of 1:3. The *amphibole group* consists of double chains with a silicon to oxygen ratio of 4:11.

The *phyllosilicates*, or sheet silicates, are composed of tetrahedra arranged in sheets (Figure 5.1d). Micas and clay minerals possess this kind of structure, with a silicon to oxygen ratio of 2:5.

The *tectosilicates* or framework silicates form a three-dimensional tetrahedra network with a silicon to oxygen ratio of 1:2 (Figure 5.2). Examples of tectosilicate minerals include the quartz (SiO_2), and the feldspar group.

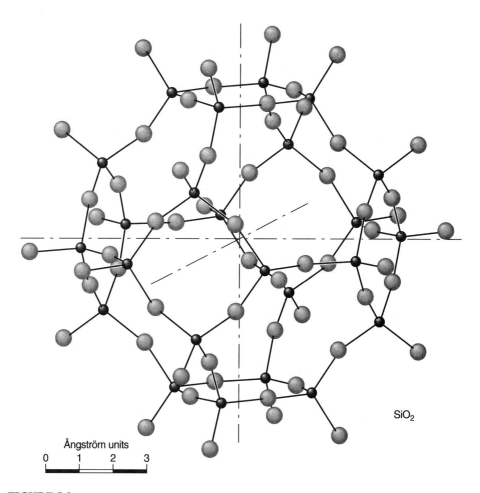

FIGURE 5.2
Linkage of silicon-oxygen tetrahedra in a three-dimensional network, typical of tectosilicates.
Source: Berry, L.G., Mason, B., and R.V. Dietrich (1983). *Mineralogy*. 2nd ed., San Francisco, CA: W.H. Freeman. Fig. 15–5, p. 388.

5.2 THE NESOSILICATES

The *olivine group* of *nesosilicates* contains two solid solution end members, *forsterite* (Mg_2SiO_4) and *fayalite* (Fe_2SiO_4). An *end member* is one of several extremes in composition. Figure 5.3 illustrates a phase diagram showing the temperature and compositional relations within olivine group minerals. *Phase diagrams* are graphs showing the stability ranges of minerals and mineral groups, as well as their chemical composition. The line labeled *liquidus* separates the field in which the system consists entirely of liquid from that which is a mixture of melt and crystals. The line labeled *solidus* is the locus of points below which the system consists completely of crystalline solids. Note that the Mg end member is a solid at all temperatures up to 1890°C, but the Fe end member is liquid above temperatures of 1205°C. Thus the Mg end member, fayalite, is much more *refractory* (resistant to melting at high temperatures) than fayalite. Olivine has a composition ranging between Mg_2SiO_4 and Fe_2SiO_4. It is a solid state mixture of these two end members, a phenomenon known as *solid solution*.

The diagnostic physical properties of the *olivine group* include vitreous luster, granular texture, and a green color for the Mg-rich varieties, which are more common than the Fe-rich, brown olivine. A transparent green variety of olivine forms the gemstone peridot, birthstone for the month of August. Forsterite is used as refractory sand.

The *garnet group* is divided into two series, each consisting of three different end members. The name of the *pyralspite group* is constructed from contraction of the first two letters of each of the three members of this group: PYrope, ALmandite, and SPessartite. The *ugrandite group* includes Uvarovite, GRossularite, and ANDradite. The chemical compositions and typical colors of each garnet species are given in Table 5.1. As a group, the garnet minerals exhibit a vitreous to resinous luster, are almost as hard as or harder than quartz (H = 6.5 to 7.5), and usually form dodeca-

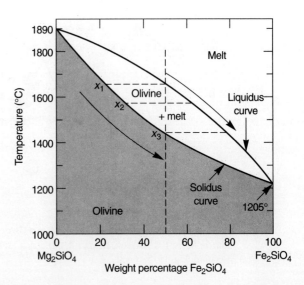

FIGURE 5.3
Phase diagram of the olivine system.
Source: Klein, C., and C.S. Hurlbut (1985). *Manual of Mineralogy*. 20th ed. New York: John Wiley & Sons. Figure 11.8, p. 374.

FIGURE 5.4
Garnet dodecahedrons, both approx. 1.5 cm diam.

hedrons or trapezohedrons (Figure 5.4). Garnet is found in metamorphic rocks. It is a relatively inexpensive gemstone, the birthstone for January, and is used extensively as a low-grade abrasive.

The mineral *zircon* ($ZrSiO_4$) is a common accessory mineral in silicic igneous rocks such as granites and granodiorites. It generally forms euhedral tetragonal crystals, but these are so small that they are usually missed in hand-specimen exami-

FIGURE 5.5
Phase diagram for the Al_2SiO_5 group.
Source: Klein, C., and C.S. Hurlbut (1985). *Manual of Mineralogy*. 20th ed. New York: John Wiley & Sons. Fig. 11.15, p. 379.

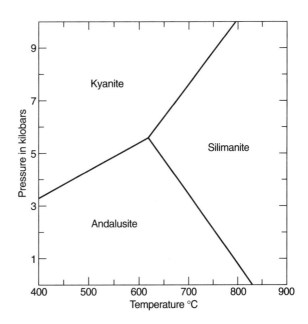

nation of most rock samples. Zircon is mined from beach sands where it may be concentrated due to its great hardness (7.5) and high specific gravity (4.6 g/cc). It is used as a source of zirconium metal, and is the birthstone for December. It is also an important mineral in radiometric age determination of rocks because it contains small amounts of U and Th.

TABLE 5.1
The Nesosilicate and Sorosilicate Minerals

Mineral Name	Chemical Composition	Hardness	Specific Gravity	Uses	Diagnostic Properties
Nesosilicates					
Andalusite	Al_2SiO_5	7.5	3.1	Refractory	Nearly square cross-section, vitreous, various shades of red, brown or green
GARNET GROUP				Gemstones, abrasives applications	Forms dodecahedrons and trapezohedrons, vitreous to resinous luster. Almandine and pyrope are red or brown, also spessartite. Andradite is various shades of green, as is grossularite. Uvarovite is emerald-green.
Almandine	$Fe_3Al_2Si_3O_{12}$	7	3.5-4.0		
Andradite	$Ca_3FeSi_3O_{12}$	7	3.5-4.0		
Grossularite	$Ca_3Al_2Si_3O_{12}$	7	3.5-4.0		
Pyrope	$Mg_3Al_2Si_3O_{12}$	7	3.5-4.0		
Spessartite	$Mn_3Al_2Si_3O_{12}$	7	3.5-4.0		
Uvarovite	$Ca_3Cr_2Si_3O_{12}$	7	3.5-4.0		
Kyanite	Al_2SiO_5	5-7	3.5	Used in refrac. porcelain	In blue bladed or tabular crystals, vitreous to silky luster.
Olivine	$(Mg, Fe)_2SiO_4$	6.5	3.5	Refractory sand	Usually as granular masses of green, vitreous crystals. Fe-rich olivine is brown.
Sillimanite	Al_2SiO_5	6	3.2		As fibrous or acicular crystals, white to gray, vitreous luster.

The Al_2SiO_5 group consists of three metamorphic polymorphs that form under varying pressure and temperature conditions. The phase diagram (Figure 5.5) for this group shows that *kyanite* is the high-pressure polymorph of Al_2SiO_5, and *andalusite* and *sillimanite* are low- and high-temperature forms, respectively. In a few places, such as in the vicinity of Pecos Baldy in the southern Sangre de Cristo Mountains of

TABLE 5.1
Continued

Mineral Name	Chemical Composition	Hardness	Specific Gravity	Uses	Diagnostic Properties
Sphene	$CaTiSiO_5$	6	3.5		Resinous to adamantine, brown to black wedge-shaped crystals.
Staurolite	$Fe_2Al_9Si_4O_{10}(O,OH)_2$	7	3.7	Right angle twins used as ornamental.	Reddish-brown, vitreous, twinned crystals form "X"'s or crosses.
Topaz	$Al_2SiO_4(F,OH)_2$	8	3.5	Gemstone	Colorless to yellow to honey-colored prismatic, vitreous crystals. High hardness.
Zircon	$ZrSiO_4$	7.5	4.5	Source of Zr.	Brown, adamantine, tetragonal dipyramids.
Sorosilicates					
Epidote	$Ca_2(Al,Fe)Al_2O SiO_4Si_2O_7(OH)$	6	3.2		Pistachio-green to dark green vitreous luster.
Vesuvianite	$Ca_{10}(Mg,Fe)_2Al_4 Si_9O_{34}(OH)_4$	6.5	3.4		Yellow to green or brown, resinous crystals. Commonly striated.

northern New Mexico, one rock sample can contain all three polymorphs, but this is very unusual.

The physical properties of each of the polymorphs are distinct. Andalusite is usually a hard (7.5), light-colored mineral that forms equidimensional, nearly square, prisms. Sillimanite is distinguished by its fibrous or acicular habit and light color. Sillimanite is fairly hard, but is noted for its fibrous habit. Kyanite is most easily recognized of the three: it forms tabular, blue or silvery blue crystals (Figure 5.6), and has a hardness ranging from five to seven, depending on the direction in which the mineral is scratched. Like andalusite, kyanite is mined for use in refractory porcelain.

Topaz ($Al_2Si_4(F,OH)_2$), birthstone for November, forms orthorhombic prismatic crystals that are typically clear or brown. However topaz can occur in a variety of colors including yellow, blue, green, and pink. It is characterized by a vitreous luster, high hardness (8) and above-average specific gravity (3.5 g/cc). Its primary mode of occurrence is in cavities of siliceous igneous rocks. Notable localities in North America are San Luis Potosi in Mexico, Ruby Mountain near Nathrop, Colorado, and the Thomas Range in Utah. Blue topaz occurs in Precambrian pegmatites and cavities in granites in Mason County, central Texas. In some localities topaz is recovered from stream gravels.

FIGURE 5.6
Bladed crystal of kyanite; well-developed cleavage traces perpendicular to blade axis. Crystal 10 cm length.

FIGURE 5.7
Twinned crystal of staurolite in Precambrian schist. Taos County, New Mexico. Individual crystals are 2.5 cm long.

The metamorphic mineral *staurolite* ($Al_2SiO_4(OH,F)_2$) forms prismatic, reddish-brown crystals but is most commonly recognized when forming interpenetrating twins at angles of either 60 degrees to form an "x", or 90 degrees to form a cross. By far the most common twins are "x"s. *Twinning*, the intergrowth of two or more crystals of the same mineral in a consistent, predictable manner characterizes many minerals. Staurolite occurs with other high grade metamorphic minerals such as garnet, kyanite, and sillimanite in schists, gneisses, and phyllites (Figure 5.7).

Sphene, also called titanite ($CaTiSiO_5$), is common in plutonic igneous rocks. Unlike zircon, which is more common in granites, sphene is more common in diorites where it forms small wedge–shaped crystals. It is found in greater abundance and as larger crystals in some metamorphic deposits. Sphene has a resinous to adamantine luster, moderate hardness (5), and occurs in various shades of brown. It is a source of titanium dioxide.

5.3 THE SOROSILICATES

The most widespread and abundant sorosilicate is the mineral *epidote*, a hydrous calcium-iron-aluminum silicate. Epidote is the most common mineral of the *epidote group*, which includes the minerals zoisite and piemontite. It forms in several metamorphic environments including hydrothermal alteration. In this latter occurrence it is a common fracture-fill of igneous rocks. In fine-grained aggregates it has an apple-green color which best distinguishes it from olive-green, magnesium-rich

FIGURE 5.8
Crystals of vesuvianite exhibiting the prism and tetragonal pyramid faces typical for this mineral.
Source: Courtesy of the Department of Library Services, American Museum of Natural History, Negative No. 2A1196.

olivine. In crystals such as those occuring in contact metamorphic deposits like the Calumet Mine in Chaffee County, Colorado, epidote is dark green, and has a hardness of six to seven. It also occurs rarely as a magmatic mineral, for example, in the dacite dikes of Boulder County, Colorado.

Vesuvianite, also called *idocrase*, is a hydrous calcium-iron magnesium-aluminum silicate forming prismatic tetragonal crystals. The name *vesuvianite* is derived from its

TABLE 5.2
The Cyclosilicate Minerals

Mineral Name	Chemical Composition	Hardness	Specific Gravity	Uses	Diagnostic Properties
Beryl	$Be3Al_2(Si_6O_{18})$	8	2.7	Gemstone, Source of Be	Hexagonal, prismatic crystals, commonly tapering, in granites and pegmatites.
Cordierite	$(Mg,Fe)_2Al_4Si_5O_{18} \cdot nH_2O$	7	2.5	Gemstone	Bluish gray vitreous crystals in metamorphic rocks.
Tourmaline	$(Na,Ca,Li,Mg,Al)(Fe,Mn)_6B_3O_9Si_6O_{18}(OH)_4$	7.5	3.1	Gemstone	Hexagonal prismatic crystals with a rounded triangular x-section, striated.

original discovery in blocks of carbonate rocks metamorphosed by the ancient lavas of Mt. Vesuvius. It is typically green or yellow, and relatively hard (6.5). Most vesuvianite crystals are vertically striated. This property and its crystal form (Figure 5.8) are diagnostic.

5.4 THE CYCLOSILICATES

This group of silicates includes the well-known minerals, beryl and tourmaline, and some less common minerals such as cordierite and axinite (Table 5.2).

Cordierite is a hydrous magnesium-iron-aluminum silicate found in gneisses and schists associated with garnet and sillimanite. It also forms in contact metamorphic environments. In a hand-specimen, cordierite resembles a grayish or bluish lump of quartz; however, unlike quartz, it is commonly altered to mica or talc. Cordierite has a hardness of 7 to 7.5 and a low specific gravity (2.6). Transparent varieties form a gemstone known as dichroite.

Beryl is a hard (8) mineral that exhibits vitreous luster and forms hexagonal prisms. Most beryl has a nontransparent bluish–green to white color, but transparent forms of beryl are prized gemstones. The most famous of these are the deep green *emeralds*, the pale blue, transparent variety known as *aquamarine*, and the rose-colored variety called *morganite*. A golden yellow variety is appropriately called *golden beryl*. The world's highest quality emeralds have traditionally come from Colombia, South America. Beryl occurs in granitic rocks, pegmatites, and metamorphic schists.

FIGURE 5.9
Single tourmaline crystal showing typical rounded triangular cross-section and vertically striated sides.

FIGURE 5.10
Black tourmaline crystals in metamorphic schist. Mora County, New Mexico. Crystals 2–3 cm long.

Common beryl is used as a source of beryllium metal, which is typically alloyed with copper.

The cyclosilicate *tourmaline* is a chemically complex hydrous silicate which for many years defied structural determination. The chemical complexity (Table 5.2) occurs because of the many substitutions that can occur in its cation sites. Tourmaline forms hexagonal crystals, commonly a trigonal, vertically striated prism (Figure 5.9). The common color is black (schorl), but gem varieties may be green (verdelite), yellow, cranberry (rubellite), or blue (indicolite). Single crystals may be zoned from pink at one end to green at the other, forming the so-called 'watermelon' tourmaline. Tourmaline occurs in metamorphic rocks (Figure 5.10), granites, and granitic pegmatites. Several districts in San Diego County, California, are famous for their high-quality tourmaline crystals.

5.5 THE INOSILICATES

The *inosilicates group* includes two important mineral subgroups: the single chain pyroxene group and the double-chain amphibole group. Both groups are chemically complex due to many solid solutions and cation substitutions.

The Pyroxene Group

As a group, the *pyroxenes* are mostly dark-colored (black to brown and green), have a moderate specific gravity (3 to 3.5 g/cc), are fairly hard (5.5 to 7), and possess two directions of cleavage at nearly right angles. Most of the chemical variation of pyroxenes can be depicted on a triangular diagram (Figure 5.11). The monoclinic pyroxenes (*clinopyroxenes*) occupy the Ca-rich portion of the quadrilateral, including the most abundant pyroxene, *augite*. The orthorhombic pyroxenes (*orthopyroxenes*) plot in the Ca-poor edge of the diagram.

The diopside-augite-hedenbergite solid solution series is common in metamorphic rocks. Diopside is green, augite is black, hedenbergite is brownish. In addition, augite is common in dark-colored igneous rocks. Chemical compositions are variable (Table 5.3).

The monoclinic pyroxene pigeonite is found in generally microscopic crystals in basaltic volcanic rocks. Orthorhombic pyroxenes form a solid solution series from magnesium-rich enstatite to iron-rich pyroxenes. Some varieties are characterized by a bronze color and unusual luster that seems to emanate from within. The orthopyroxenes are common in plutonic rocks such as peridotites, pyroxenites and gabbros.

Normally a cream-colored mineral, spodumene ($LiAlSi_2O_6$) is an exception to the typical dark color of pyroxenes. The two gem varieties of spodumene are the lilac-colored *kunzite*, and emerald green *hiddenite*. Spodumene is characteristic of lithium-rich pegmatites such as the Harding Pegmatite near Dixon, New Mexico (Figure 5.12) and the many pegmatites of the Harney Peak Granite of the Black Hills, South Dakota. It is a source of lithium, which is increasingly used in batteries, as a grease additive, and as a medication for manic-depressive illness.

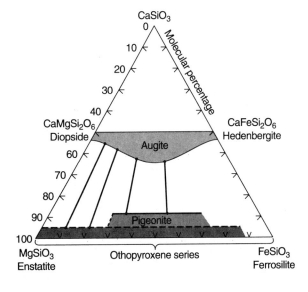

FIGURE 5.11
Triangular diagram showing pyroxene compositions.
Source: Klein, C., and C.S. Hurlbut. (1985). *Manual of Mineralogy*. 20th ed. New York: John Wiley & Sons. Fig. 11.40, p. 398.

TABLE 5.3
The Inosilicate Minerals

Mineral Name	Chemical Composition	Hardness	Specific Gravity	Uses	Diagnostic Properties
Pyroxene Group					
Aegirine	$NaFeSi_2O_6$	6.5	3.5		Acicular or prismatic, dark green crystals, vitreous, occurs in alkalic rocks.
Augite	$(Ca,Na,Mg,Fe,Al)(Si,Al)_2O_6$	6	3.2		Prismatic, black, vitreous crystals showing a nearly square x-section.
Enstatite	$MgSiO_3$	6	3.2		Greenish yellow to brown, vitreous to pearly luster, usually granular.
Diopside	$CaMgSi_2O_6$	6	3.2		Light green prismatic crystals, common in contact metamorphic rocks.
Jadeite	$NaAlSi_2O_6$	6.5	3.4	A form of jade	Apple-green, compact masses.
Spodumene	$LiAlSi_2O_6$	6.5	3.2	Source of Li, gemstones	White to gray, vitreous, striated bladed crystals. Found in pegmatites.
Amphibole Group					
Actinolite	$Ca_2(Mg,Fe)_5Si_8O_{22}(OH)_2$	5.5	3.2		Green aggregates of vitreous acicular of fibrous crystals.
Anthophyllite	$(Mg,Fe)_7Si_8O_{22}(OH)_2$	6	3.0		Brown or greenish brown, fibrous or acicular crystals.

TABLE 5.3
Continued

Mineral Name	Chemical Composition	Hardness	Specific Gravity	Uses	Diagnostic Properties
Cummingtonite	$(Mg,Fe)_7Si_8O_{22}(OH)_2$	6	3.3	Minor source of asbestos	Light brown, vitreous to silky crystals.
Glaucophane	$Na_2Mg_3Al_2Si_8O_{22}(OH)_2$	6	3.3		Blue to purple vitreous crystals. Occurs in blueschists.
Hornblende	$(Ca,Na)_2(Mg,Fe,Al)_5Si_6(Si,Al)_2O_{22}(OH)_2$	6	3.2		Black to greenish black, prismatic or acicular crystals.
Riebeckite	$Na_2Fe_5Si_8O_{22}(OH)_2$	6	3.3	Minor source of asbestos	Blue to black, acicular radiating crystals, vitreous, occurs in alkalic rocks.
Tremolite	$Ca_2Mg_5Si_8O_{22}(OH)_2$			Minor source of asbestos	White to gray, acicular, radiating crystals. Occurs in metamorphosed limestones.
Pyroxenoid Group					
Pectolite	$Ca_2NaHSi_3O_9$	5	2.8		White, acicular crystals, in radiating masses vugs basalt.
Rhodonite	$MnSiO_3$	6	3.5	Ornamental stone	Pink or pinkish-brown masses associated with other Mn minerals.
Wollastonite	$CaSiO_3$	5	2.8	Minor use in ceramics	White or gray tabular crystals, vitreous. Occurs in metamorphosed limestones.

FIGURE 5.12
Bladed crystals of spodumene bounding sample of Harding pegmatite. Taos County, New Mexico. Blades of spodumene approx. 25 cm long.

Another group of pyroxenes, not represented in Figure 5.11, are the sodium pyroxenes jadeite ($NaAlSi_2O_6$) and aegerine ($NaFeSi_2O_6$). Their chief distinguishing characteristic is a green or greenish-brown color. Although *jadeite* is strictly a metamorphic mineral, occuring in association with minerals like glaucophane and lawsonite, *aegerine* occurs in igneous and metamorphic rocks.

Jadeite is one of the mineral varieties known as jade. The other variety of jade is *nephrite,* an amphibole. For centuries Chinese have sought jade as a prized material for carving delicate art objects.

The Amphibole Group

The *amphibole group* in general displays a greater diversity of color than the pyroxenes, generally occuring in shades of gray, green, or brown. Amphiboles (Table 5.3) tend to be softer (5.5 to 6) than the pyroxenes, and have a specific gravity of about 3 to 3.5 g/cc, similar to that of pyroxenes.

The solid solution series tremolite-actinolite ranges in color from green actinolite to white or gray tremolite. Both end–members occur in acicular or fibrous forms, generally as radiating aggregates (Figure 5.13). Both are common in metamorphic environments.

The most common amphibole, *hornblende*, is characteristically black and identified by two directions of cleavage at angles of about 60 and 120 degrees. It occurs in volcanic and plutonic rocks, as well as metamorphic rocks, where it is a common, abundant constituent in amphibolite.

Glaucophane and riebeckite are two bluish sodium-bearing amphiboles. *Glaucophane* is exclusively metamorphic but *riebeckite* occurs in both igneous and metamorphic rocks. *Crocidolite* is the fibrous variety of riebeckite and the rarer but toxic variety of asbestos. The silicified form of crocidolite, colored yellow or brown by a small amount of oxidized iron, forms an attractive irridescent rock known as *tiger's-eye*.

The Pyroxenoid Group

The *pyroxenoid group* (Table 5.3) contains a common and abundant metamorphic mineral, wollastonite, and several less well-known minerals. Their structure consists of twisted single chains of tetrahedra yielding a triclinic crystal structure.

FIGURE 5.13
Radiating acicular crystals of actinolite. Calumet Mine, Chaffee County, Colorado. Individual crystals 3 cm long.

FIGURE 5.14
Chrysotile; fibrous variety of serpentine and major form of asbestos. Fibers are 2 cm long.

Wollastonite ($CaSiO_3$) is a white or gray fibrous mineral, and a common constituent of impure limestones altered by contact metamorphism. *Pectolite's* physical properties closely resemble those of wollastonite, but it occurs in the vugs of basaltic rocks. Another mineral having the same mode of occurrence and similar composition as pectolite is scolecite, a zeolite group mineral (Figure 5.21).

The mineral *rhodonite* ($MnSiO_3$) is characterized by its pink color. In this respect it resembles another manganese mineral, rhodocrosite ($MnCO_3$). Rhodonite is distinguished from rhodocrosite by its greater hardness and lack of rhombohedral cleavage.

5.6 THE PHYLLOSILICATES

The layered silicate structures defining the group known as *phyllosilicates* include the serpentine, clay mineral, and mica subgroups. Each subgroup is discussed separately in the following paragraphs.

The Serpentine Subgroup

The *serpentine* subgroup includes three members of similar chemical composition but distinct habits. The layered or platy variety of serpentine, *antigorite*, usually exhibits a green to yellowish-green color and waxy luster. The fibrous variety, *chrysotile*, usually exhibits a silky luster and various shades of grayish-green.

Lizardite is a massive variety of serpentine. Serpentine minerals are of metamorphic origin. The fibrous variety is the major form of asbestos (Figure 5.14) and usually occurs in veins or layers within which the fibers are oriented perpendicular to the layering. Asbestos fibers have many uses resulting from their flexibility, which allows them to be woven, and their insulating and heat-resistant properties. However, chrysotile is a carcinogen if inhaled over long periods of time.

The Mica Subgroup

The *mica* group is a welcome change for beginning mineralogy students who may suspect that the number of mineral species is almost endless and their physical properties too similar. Though chemically complex (Table 5.4), micas are readily identified by their micaceous cleavage (perfect in one direction) and colors.

The greenish to yellowish-white mica, *muscovite* (Figure 5.15), is so-called because it was used as a substitute for window glass in Russia and later found widespread use as "windows" for wood-burning stoves. More recently it has been used as electrical insulating material and is still commonly seen as windows in floor furnaces. The electrical applications, however, are being gradually phased out and replaced by ceramic materials.

FIGURE 5.15
Small "book" of muscovite; showing the pseudohexagonal form of the micas. Book is 4 cm long.

TABLE 5.4
The Phyllosilicate Minerals

Mineral Name	Chemical Composition	Hardness	Specific Gravity	Uses	Diagnostic Properties
Biotite	$K(Mg,Fe)_3AlSi_3O_{10}(OH)_2$	3	3.0		Dark brown to black mica. Thin sheets flexible, forms tabular pseudohexagonal crystals.
Chlorite	$(Mg,Fe)_3(Si,Al)_4(OH)_2O_{10}$	2	3.0		Dark green mica, vitreous luster. Common in metamorphic rocks.
Chrysocolla	$Cu_4H_4Si_4O_{10}(OH)_8$	3	2.4	Ornamental	Bluish green, compact crpytocrystalline masses.
Kaolinite	$Al_2Si_2O_5(OH)_4$	2	2.5	Porcelain, pottery and china	White, very fine grained masses with earthy luster, soft.
Lepidolite	$K(Li,Al)_3Si_3O_{10}(O,OH,F)_2$	3	2.8	Source of Li.	Pink to lilac-colored mica, found in pegmatites with other Li minerals.
Muscovite	$KAl_3Si_3O_{10}(OH)_2$	2.5	2.8	Formerly electrical insulator	Clear mica in thin sheets. Thin sheets flexible and elastic.
Phlogopite	$KMg_3AlSi_3O_{10}(OH)_2$	3	2.9	Formerly electrical insulator	Yellowish-brown or gold colored mica. Occurs in ultramafic rocks.
Pyrophyllite	$Al_2Si_4O_{10}(OH)_2$	1.5	2.8		White, green, gray or brown radiating aggregates of platy crystals.
Serpentine	$Mg_3Si_2O_5(OH)_4$	4	2.5	Asbestos	The variety antigorite is a green, massive, waxy-like mineral. The chrysotile variety, is fibrous, light colored, with silky luster. Major source of asbestos.

Large crystals, or "books" of muscovite (Figure 5.15), occur in granitic pegmatites; muscovite is a common mineral in granites. Muscovite is also abundant in schists, gneisses, and phyllites.

The greenish-black mica, *biotite*, is actually dark brown in thin sheets. Like muscovite, it commonly forms hexagonal-looking, six-sided crystals or "books," but actually has monoclinic symmetry. Biotite is less common in granitic pegmatites, but is a common accessory mineral in plutonic, volcanic, and metamorphic rocks.

The gold mica, *phlogopite*, has a golden luster in thick sheets. In thin sheets it is light yellow and may be easily confused with biotite. However, it is not nearly as abundant as biotite. Phlogopite occurs in a number of unusual environments: in association with ultramafic (Fe- and Mg-rich rocks; e.g., peridotite) rocks, as an abundant component in kimberlites, and in some metamorphic deposits.

The pink mica, *lepidolite*, has a soft pearly luster and shades of color from pink to lilac. Its occurrence is restricted to pegmatites where it is associated with other lithium minerals such as spodumene. Lepidolite is found in pegmatites in the Black Hills, South Dakota; New England; and the Harding Pegmatite of northern New Mexico.

Chlorite is a green mica-like mineral forming a separate group distinct from the micas. Formed by alteration and metamorphism, it is widespread but rarely forms large crystals. It is a major component and "coloring agent" of greenschist, a rock type resulting from low-grade regional metamorphism.

The Clay Mineral Subgroup

Not until the 1960s, when X-ray diffraction analysis of mineral structure and composition became widespread, did this group become delineated. "Clay," in layman's terms, usually refers to a fine-grained material with earthy luster. It usually consists of one or more distinct *clay minerals*. Clay minerals are subdivided into the *kaolinite group* and the *smectite group* by the way in which the layers of silicate tetrahedra are arranged.

Kaolinite ($Al_2SiO_5(OH)_4$) is the only abundant member of the kaolinite group. It is earthy and usually white, unless colored by iron or other impurities. Kaolinite forms by weathering or alteration of feldspars and is one of the most useful of clay minerals. The higher grade clays, relatively free of impurities that can color the final product, are white and used in making fine china, ceramics and pottery, and as a filler in papers. Lower grade clays are used as a major constituent of bricks, tiles, and a wide range of pottery products such as flower pots.

The smectite group includes a subgroup characterized by their ability to absorb water. These are known as the *swelling clays*—a property both useful and destructive. *Montmorillonite* is the common member of this group, occurring in many shales and claystones. Buildings, homes, and highways built on this clay are subject to warping due to the clay's expansion and contraction. Complaints of buckled sidewalks, cracked foundations, doors that bind, and windows that stick are common among those who have built homes on montmorillonite clay. A variety of swelling clay known as *bentonite* originates from altered volcanic deposits. It is mined in Wyoming and other places for the manufacture of drilling muds.

5.7 THE TECTOSILICATES

The quartz (silica) and feldspar subgroups are the most common members of the tectosilicate group (Table 5.5). Other tectosilicates are the feldspathoids and the zeolites.

The Silica Subgroup

The *silica* subgroup contains eight polymorphs, the greatest number known in the mineral realm—although only quartz is commonly known. Less widely known, but nonetheless very abundant, are the high-temperature minerals *cristobalite* and *tridymite*. In addition, quartz occurs in several microcrystalline forms that are known by a variety of names, including flint, chert, and agate. The natural hydrated form of silica is called *opal* ($SiO_2 \cdot H_2O$).

The mineral quartz occurs in a low-temperature (α) 0°C–550°C form and a high-temperature (ß) 850°C–1750°C form (Figure 5.16). The low-temperature form is common and readily recognized by its hexagonal prism terminating in two sets of rhombohedrons that combine to form a hexagonal-like pyramid (Figure 5.17). The

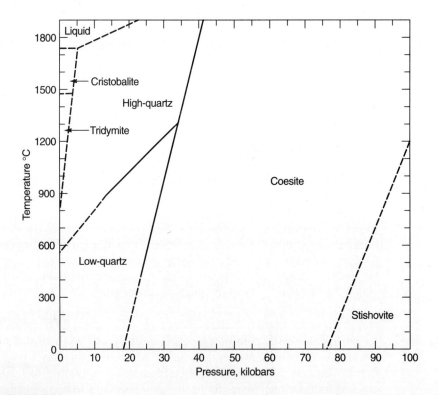

FIGURE 5.16
Phase diagram for SiO_2 polymorphs.
Source: Klein, C., and C.S. Hurlbut (1985). *Manual of Mineralogy*. 20th ed. New York: John Wiley & Sons. Fig. 11.87, p. 440.

FIGURE 5.17
Quartz crystal showing typical hexagonal prism crowned by two sets of rhombohedra combined to form apparent hexagonal pyramid.

FIGURE 5.18
Fragment of quartz crystal showing characteristic conchoidal fracture.

TABLE 5.5
The Tectosilicate Minerals

Mineral Name	Chemical Composition	Hardness	Specific Gravity	Uses	Diagnostic Properties
Silica Subgroup					
Cristobalite	SiO_2	6.5	2.3		Generally white or gray spherical aggregates in volcanic glass.
Opal	$SiO_2 \cdot nH_2O$	6	2.2	Gemstone	Common opal is vitreous to waxy. Precious opal shows a play of colors.
Quartz	SiO_2	7	2.7	Gemstones, in china and glass	Occurs in many varieties; both crystalline and microcrystalline forms.
Tridymite	SiO_2	7	2.2		Small crystals in volcanic rocks.
Feldspar Subgroup Alkali Feldspars					
Microcline	$KAlSi_3O_8$	6	2.6	Used in porcelain	Vitreous white to brown, two cleavage planes at 90° angles. Amazonite is blue or green.
Orthoclase	$KAlSi_3O_8$	6	2.6	Same as Microcline	White, red or brown, vitreous, 2 cleavage directions at 90° angles.
Sanidine	$KalSi_3O_8$	6	2.6		Colorless, vitreous crystals. Forms phenocrysts in volcanic rocks.

TABLE 5.5
Continued

Mineral Name	Chemical Composition	Hardness	Specific Gravity	Uses	Diagnostic Properties
Plagioclase Feldspars					
Albite	$NaAlSi_3O_8$	6	2.6	Used in Porcelain	Colorless to light gray. Vitreous. Shows albite twinning.
Anorthite	$CaAl_2Si_2O_8$	6	2.7	Ornamental stone	Dark or bluish gray, vitreous. Shows albite twinning.
Feldspathoid Subgroup					
Leucite	$KAlSi_2O_6$	6	2.5		Light gray, trapezohedral crystals
Nepheline	$NaAlSiO_4$	6	2.6	Used in ceramics	Vitreous to greasy luster, colorless to light brown. Occurs in alkalic rocks.
Zeolite Subgroup					
Analcime	$NaAlSi_2O_6 \cdot H_2O$	5.5	2.2	Molecular sieve	White, cubic, vitreous crystals. Occurs in basalts and basalt cavities.
Heulandite	$CaAl_2Si_7O_{18} \cdot 6H_2O$	4	2.2	Molecular sieve	White or yellow vitreous crystals with pearly luster
Scolecite	$CaAl_2Si_3O_{10} \cdot 3H_2O$	4	2.2	Molecular sieve	White or light gray, acicular radiating crystals in basalt vugs.
Stilbite	$CaAl_2Si_7O_{18} \cdot 7H_2O$	4	2.2	Molecular sieve	Sheaflike groups of white, vitreous to pearly crystals in basalt vugs.

prism and rhombohedral faces are not usually of equal size. Quartz does not have cleavage and breaks with a conchoidal fracture like that of glass (Figure 5.18). Low-temperature quartz forms in a wide spectrum of environments, including sedimentary, metamorphic, and plutonic, as well as hydrothermal ore deposits and pegmatites. Its presence in sedimentary rocks is due both to its resistance to physical and chemical weathering and its formation in sedimentary environments. High-temperature quartz occurs in silisic volcanic rocks. Its crystals are usually doubly terminated, consisting of hexagonal dipyramids. The transition from high to low quartz is nearly instantaneous because it involves minor internal atomic structural changes; however, rapid cooling in volcanic rock may permit retention of the high-temperature crystal form.

Transparent, well-formed quartz crystals occur in a variety of colors and with various elemental inclusions. Pink *rose quartz* rarely forms crystals and then only small ones; *smoky quartz* is dark gray to black, its smoky color probably resulting from natural radiation; *citrine* is a light yellow, gem-quality quartz; *amethyst's* beautiful deep purple color is produced by traces of iron. Needles of rutile in attractive, yellow to brown radiating arrays are commonly found within transparent quartz crystals. This so-called *rutilated quartz* is an attractive and inexpensive gemstone. Rarely, needles of zircon grow within quartz: the dispersion that these small crystals produce gives the quartz a bluish color.

In addition to the many varieties of macrocrystalline SiO_2 there are a large number of fine-grained (microcrystalline) forms. They are subdivided into fibrous (*chalcedony*) or granular (*chert*) varieties on the basis of their microscopic characteristics. Chalcedony has a soft pearly luster, is commonly banded, and occurs in a variety of colors—most often a shade of blue or gray. *Carnelian* is a red variety of chalcedony. *Agate* is another variety of chalcedony, typically occuring as cavity fillings having concentric or layered patterns of different colors (Figure 5.19). The many names for different types of agate are based on color and the internal layering patterns. *Jasper* is an opaque form of chalcedony, usually red or brown. Chert or flint is a fine-grained variety of quartz that breaks with a smooth conchoidal fracture. Chert is normally found in a variety of colors (flint usually refering to black or dark-colored chert), and occurs as *nodules* in carbonate sedimentary rocks such as limestone.

Opal is a gemstone, the quality of which varies with its brilliance and display of colors. The color variety is due to the refraction or splitting of the visible light spectrum into a range of colors by its hydrated inclusions. A variety of names have been given to the gem varieties of opal—Precious Opal, Black Opal, and Fire Opal. Fire opal has an internal flash of colors dominated by reds and oranges. Common opals lack the fire of the gem varieties, and are distinguished from microcrystalline quartz varieties by their resinous luster and lower specific gravity. Opals occur principally in the cavities of volcanic rocks and as veins and pockets in sandstones. Classic precious opal localities are Queretaro, Mexico, and the sandstones of Coober Pedy in southern Australia.

The high-pressure polymorphs of silica are *coesite* and *stishovite*. As would be expected of such high-pressure forms, they have the highest specific gravity of all the

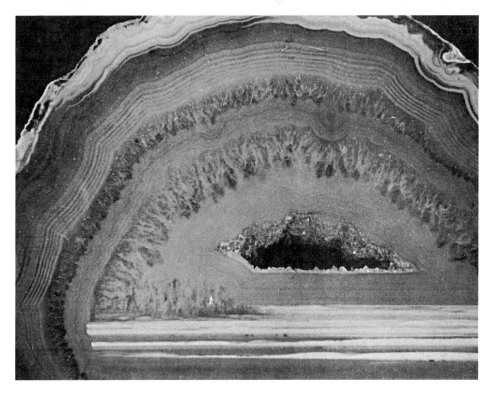

FIGURE 5.19
Slab of polished agate showing fine concentric banding. Slab is 10 cm in diameter.

polymorphs (3 to 4.3 g/cc), and have been found only in association with meteorite and asteroid impact craters, where extremely high pressures have been produced during impact with silica-bearing rock. The polymorphs have also been artificially created by the pressures generated by below-ground and surface nuclear weapons test blasts. Coesite has also been found in high-pressure igneous and metamorphic rocks.

The Feldspar Subgroup

The composition relations of the alkali and plagioclase series are shown in Figure 5.20. Complete solid solution exists in the plagioclase series from *anorthite* ($CaAlSi_2O_8$) to *albite* ($NaAlSi_3O_8$) for those feldspars formed in plutonic rocks. The alkali feldspar series shows complete solid solution only at higher, volcanic temperatures (900°C to 1200°C). Note that the albite end member occurs in both series.

The members of the plagioclase feldspar series show gradation from light-colored albite to dark gray anorthite. Feldspars have two directions of good cleavage, at nearly right angles. Plagioclase's specific gravity ranges from 2.6 g/cc for albite, to

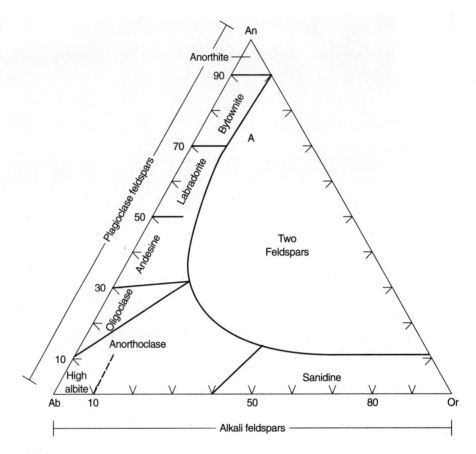

FIGURE 5.20
Feldspar classification triangle showing alkali- and plagioclase-feldspar series.
Source: Deer, W.A., R.A. Howie, and J. Zussman (1963). *The Rock Forming Minerals*. London: Longmans, Green and Co. v. 4, Fig. 1, p. 2.

2.75 g/cc for anorthite. The most reliable distinguishing characteristic of the plagioclase series is the presence of polysynthetic (multiple) twinning. In hand-specimens and under the magnifying lens, these twin planes are expressed as parallel striations on crystal faces and cleavage surfaces.

The alkali feldspar series includes several distinct species—*microcline, orthoclase, sanidine* and *anorthoclase*—but the distinctions between these forms are not easily made. All alkali feldspars are distinguished from plagioclase by an absence of twinning striations. Microcline and orthoclase occur in plutonic and metamorphic rocks. Sanidine and anorthoclase are the alkali feldspars found in volcanic rocks. A variety of microcline, *amazonstone*, occurs in pegmatites and cavities in granites, and has a distinctive turquoise-blue color. Orthoclase is distinguished from the other feldspars by the 90-degree angle between its two good cleavage directions.

The Feldspathoid Subgroup

Feldspathoid minerals are chemically and structurally similar to the feldspars but comparatively deficient in silica (Table 5.5). The two most common feldspathoids are *leucite* and *nepheline*. Leucite ($KAlSi_2O_6$) is found in alkalic rocks such as the lava flows of the Leucite Hills, Wyoming. It is characterized by light-colored, equidimensional, trapezohedral crystals. Nepheline ($NaAlSiO_4$) occurs in alkalic intrusive (plutonic) and extrusive (volcanic) rocks. It is common in syenites, alkalic pegmatites, and an extrusive rock known as phonolite. Nepheline resembles quartz in hand-specimen but is softer and has a resinous luster.

The Zeolite Subgroup

Zeolites comprise more than 15 minerals, only a few of which are common. They occur in volcanic, volcaniclastic (explosively fragmented volcanic rock), and sedimentary rocks. Originally thought to occur mainly in vugs of volcanic rocks, we now know that zeolites are a principal component of many altered tuffs (fine-grained volcaniclastic rocks) and rhyolites.

Zeolites have a long, impressive list of properties making them suitable for a large number of process applications. Their internal crystalline structure allows them

FIGURE 5.21
Radiating aggregate of scolecite crystals from a vug in a basalt. Aggregate is 3 cm in diameter.

to function as "molecular sieves." They also serve in dehydration, absorption, and ion exchange reactions. The uses of natural zeolites are many, including soil applications where they serve as aerators, as neutralizers of acidic soils, and regulators of ammonia, nitrogen, and potassium release from fertilizers. In this latter application zeolites act as "time-release capsules." Other applications include carriers of trace elements in animal feed, deodorizers that make some wastes acceptable cattle feed, and for controlling toxic ammonia in fish farms. In the future zeolites may serve as radionuclide "sponges." The zeolite clinoptilolite selectively absorbs radioactive cesium and strontium from low-level nuclear waste.

Most zeolites are fine grained, and only a few species crystallize in cavities as recognizable crystals. Most of their crystal habits are fibrous (Figure 5.21), including the minerals natrolite, scolecite, and stilbite. They are typically colorless or white, with a hardness between four and six, and a specific gravity less than 2.5 g/cc. Other zeolites forming macroscopic crystals are heulandite (monoclinic) and chabazite (hexagonal).

6
The Structure and Composition of the Earth

Before resuming our journey through the wonders of the world of rocks and minerals of the crust, let us examine what we we know about Earth's interior. The Earth consists of a relatively thin (10 to 60 km) outer layer called the *crust*, an intermediate layer called the *mantle*, and a central portion called the *core*. The mantle and core are more inaccessible than the Moon's surface. We may drill to the top of the mantle in the near future, but the lower mantle and core will remain unreachable. Nonetheless, specimens of the upper part of the mantle are brought to the surface in kimberlite pipes and some basalt flows. Let us begin our discussion of Earth's interior by listing the types of data available and how these provide clues to the internal structure and composition of the Earth.

6.1 SOURCES OF INFORMATION

The first piece of evidence is the size and mass of the Earth. Approximately 8,000 miles in diameter, Earth is calculated to have a mass of 6×10^{24} kilograms. These numbers can be reduced to a more usable form of Earth's average density: approximately 5.5 gm/cc (Robertson, 1966).

A second source of information about Earth's interior is the behavior of seismic waves traveling within. This branch of geology, called *seismology*, is one of the best sources of information about, and techniques for, determining the internal structure of the Earth (Figure 6.1). Fragments of the upper mantle included within volcanic rocks provide additional information. This evidence suggests that the upper mantle has an olivine-pyroxene (*peridotite*-like) composition.

Experimental data on the effects of changing temperature and pressure on mineral stability, combined with model *geotherms* (graphs of the rate of change of temperature and pressure with depth), enable prediction of how stable mineral assemblage composition changes with increasing depth. These studies confirm that pyroxene and olivine are stable at the pressures of the upper mantle but change to denser forms, such as *spinel* and *perovskite*, lower in the mantle.

Another important piece of evidence yielding vital clues to the core's composition are the *meteorites*—fragments of planetary material originating elsewhere in our solar system. Meteorites have fallen to Earth, captured by its gravitational field.

Using clues provided by the information sources above, a model of Earth's interior can be constructed. Additional information can be used to test the model. In science, theories are almost never proven correct, but can be proven wrong, and are subject to constant revision as new data become available. The model that is accepted today may prove untenable or in need of revision with additional data tomorrow.

6.2 EXPLORING THE EARTH'S INTERIOR— A HISTORICAL PERSPECTIVE

The science of instrumental seismology began around the turn of the century (1900) and soon afterward, seismologists had delineated the three main subdivisions of Earth (Figure 6.1). When an earthquake or explosion occurs within or on the surface of Earth, part of the energy propagates as shock waves transmitted through the

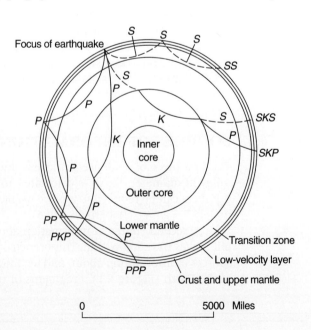

FIGURE 6.1
Major internal subdivisions of the Earth and idealized paths of seismic waves through its interior showing how the wave depth path varies with distance from the focus. Compressional waves (P, solid lines) refracted sharply by core. Shear waves (S, dashed lines) end at the core, traverse core as compressional waves, emerging in mantle as P and S waves.
Source: Robertson, E.C. 1966. *U.S. Geol. Surv. Circ.* 532, p. 8.

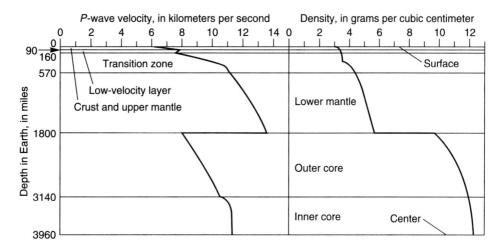

FIGURE 6.2
Siesmic wave velocity within the Earth as function of depth.
Source: Robertson, E.C. 1966. *U.S. Geol. Surv. Circ.* 532, p. 8.

rocks. The shock waves' velocity depends on the density and elasticity of the rocks through which they travel. There are two types of seismic waves: *surface waves* travel along Earth's surface or along the boundary between two layers and *body waves* travel through Earth's interior, and include *Primary* (P) and *Secondary* (S) waves. Body waves are reflected and refracted at boundaries between rock layers where seismic velocity changes. In this respect they behave as light does when it travels through a transparent mineral, obeying the laws of optics.

For a given rock type P-waves travel faster than S-waves. Because fluids have zero rigidity, the S-wave (a shear wave) cannot propagate through fluids. The P-wave is compressional, and thus, propagates through fluids, but at a reduced velocity. The path of shock waves generated by a major earthquake is shown diagramatically in Figure 6.1. The location where the rupture occurs is called the *focus*; the spot on Earth's surface directly above the focus is called the *epicenter*. The ray paths curve as wave velocity increases with depth in the same way a highly refractive crystal causes a beam of light to bend more than in a crystal with a low refractive index. A seismological observatory close to the focus will also receive earthquake waves having traveled at shallower depths than more distant observatories. At internal angles less than 103 degrees from the focus, earthquake wave velocity and strength are unaffected unless a body of magma (liquid rock) lies along the wave path. From 103 to 142 degrees from the focus, a shadow zone is present within which P- and S-waves are absent or very weak. Some weak waves arise by refractions from the mantle-core interface, or as reflections from the outer/inner core boundary. Between 142 and 180 degrees, the P-wave is present but delayed by its passage through the outer core. The S-wave is not present because it cannot traverse the outer core. Delayed P-wave arrival and S-wave absence suggest that the outer core is a fluid. So a major discontinuity (the *Gutenburg Discontinuity*) is postulated to exist at a depth of about 2,900 km between a solid mantle characterized by fast P- and S-wave velocities and an outer core with slower P-waves and no S-waves.

In 1936 it was discovered that P-waves traveling at depths greater than 5,000 kilometers increased in velocity. This behavior is consistent with, and is now interpreted as, reflecting a solid inner core. Several years later it became obvious that another layer of low velocity material was present in the upper mantle from depths of about 100 km to as much as 300 km. Called by seismologists the *low velocity zone*, or *LVZ*, it became part of the evidence that Harry Hess used to explain the movement of the continents and the basic mechanism of Plate Tectonics Theory. In the plate tectonic framework Earth's upper-most structural layer is divided into three parts. The uppermost, solid, brittle layer, including the crust and uppermost mantle, is called the *lithosphere*. The upper portion of the underlying layer characterized by low seismic wave velocities is labeled the *asthenosphere*. The *Mohorivicic Discontinuity*, or boundary between the crust and mantle, lies at depths of 10 to 60 kilometers; its minimum depth occurs in the crust of the ocean basins, whereas deeper boundaries lie beneath the crust of continents, and the greatest depths beneath the highest mountain ranges. The lithosphere is about 100 km thick. The asthenosphere is about 700 kilometers thick, the upper portion of which is interpreted as a zone in which *partial melting* (differential melting of Si- and Al-rich rock) occurs locally.

P-wave seismic velocities range from 3 to 6 km/sec in the crust. Crossing at the Mohorivicic Discontinuity into the upper mantle they increase to about 8 km/sec, and continue increasing with depth to more than 13 km/sec just above the core-mantle boundary (Figure 6.2). At the core P-wave velocity decreases to about 8 km/sec, then increases with depth to more than 11 km/sec at Earth's center (Robertson, 1966).

Temperature and pressure both increase with increasing depth. The role that each parameter plays at different depths is a function of mineralogy, but in general pressure and temperature work against each other. Increasing pressure produces a solid Earth because solid states are generally more tightly packed, and thus, more dense than are liquid states. The effect of increasing temperature is the melting of minerals. The combined effect of these parameters produces a zone of partial melting, the LVZ, and at a depth of 2,900 km the mafic silicate mantle is solid but the Fe-Ni outer core is fluid.

Fragments of Earth's interior provide another clue to the mantle's composition. These fragments, found in igneous rocks, include a wide variety of rock types. Most of these *xenoliths*—fragments of rock foreign to the body in which they are found—are pieces of crustal material, including parts of the sedimentary sequence through which the original magma intruded. However, a suite of ultramafic rocks not characteristic of crustal rocks occur in some basalts and in kimberlites. These xenoliths include *peridotite* composed of pyroxene and olivine, *pyroxenite* composed of pyroxenes, and *garnet peridotite* composed of garnet, pyroxene and olivine. Some of these xenoliths also contain diamond crystals indicating origins at several hundred kilometers depth (Meyer, 1977). In addition, but less commonly, fragments of *dunite* composed of olivine and the rock known as *eclogite*, consisting of garnet and the sodium-rich pyroxene omphacite, are found in kimberlite xenoliths. Based on this suite of xenolith compositions, the upper mantle is inferred to be composed of pyroxene and olivine + garnet and spinel to a depth of at least a few hundred kilometers. This conclusion is supported by seismic wave velocities for these rock types measured in the laboratory that are identical to those observed in the upper mantle.

The core's composition is more problematic because we have no samples of it. Our only nonexperimental evidence of the core's composition is based on seismic wave behavior, and even that view is clouded by the "haze" of the lower mantle through which these waves must travel before reaching seismographs. Thus, we need another source of information—the meteorites.

6.3 EVIDENCE FROM THE METEORITES

Meteorites are remnants of fragmented or incompletely formed planetary bodies from within our solar system that have fallen to Earth. Most meteorites are found long after reaching the surface. A large number of meteorite fragments have been recovered from the High Plains of the central United States. However, they are most commonly found in Antarctica, where the movement of glacial ice has concentrated them in certain locations.

The three principal meteorite types are:

Iron Meteorites
Stony-Iron Meteorites
Stony Meteorites

Meteorites classification beyond that of this simple grouping is useful only to specialists. The iron meteorites are the most readily identifiable because they look like lumps of rusty metal. They are composed primarily of two alloys of iron and nickel, *kamacite* and *taenite*. In those known as *octahedrites*, these two alloys form an exsolution ("segregation") pattern called the *Widmanstatten Structure* (Figure 6.3). This structure's origin is important to our understanding of meteorite formation and their genetic connection to Earth. The exsolution pattern is the result of slow cooling within a planetary body or large asteroid. During this slow cooling, plates of nickel-poor kamacite (6 percent Ni) nucleate and grow in a crystallographic orientation within a host of nickel-rich taenite (30 to 60 percent Ni). This pattern implies that slow cooling occurred in the cores of small planetary bodies that have subsequently broken up. Might the composition of these cores be similar to that of Earth's?

Stony meteorites are composed of olivine and pyroxene, a composition similar to that of Earth's mantle. The stony-iron meteorites are intermediate in composition between iron and stony meteorites, and probably formed in a transition zone between the core and mantle of the parent planetary bodies. Of the meteorite falls observed, about 90 percent are stony meteorites and about six percent are iron mete-

FIGURE 6.3
The Widmanstatten Structure in an octahedrite—the Bagdad, Arizona, iron meteorite.
Source: Moore, C. B., and P. P. Sipiera. 1975. Center for Meteorite Studies, Arizona State Univ. Pub. 13, p. 8.

orites. By comparison, Earth's mantle constitutes 84 percent of its volume and the core about 16 percent.

Other evidence strengthening the genetic link between the meteorites and Earth are their ages of formation. The age of Earth, first determined by radiometric dating in the mid-1950s (Patterson, 1956) is approximately 4.55 billion years. This age probably represents the time when the Earth coalesced into a discrete body orbiting the Sun. The other planets must also have formed at this time. Most meteorites analyzed to date give this same age—4.55 or 4.6 billion years. Thus, the evidence suggests that meteorites represent fragments of planetary bodies that are genetically, and therefore compositionally, linked to the origin of Earth and the other small, dense, and rocky planets of the inner solar system (Mars, Venus, Mercury, and Earth's Moon).

Based on meteorite evidence, a probable composition for Earth's core is an alloy of iron and nickel. This composition has the appropriate density (7 to 8 g/cc) to yield an Earth average of 5.5 g/cc; has the appropriate seismic wave velocities for its core; and Fe/Ni alloy has the proper chemical properties to produce a liquid outer core and solid inner core. The alloy also possesses the magnetic parameters necessary to produce Earth's magnetic field. This field is believed to originate in the outer fluid portion of the core by the action of currents produced by Earth's rotation and by heat convection within.

6.4 PLATE TECTONICS AND THE ORIGIN OF THE CRUST

Now the composition and origin of the crust will be addressed. Do there exist meteorites with the composition of Earth's crust? Yes, but any meteorite with such a crustal composition would likely go unnoticed. However, there exists a group of meteorites that does not fit any of the standard categories. These have mostly basaltic compositions and yield radiometric ages much younger than that of Earth's origin. These *shergottites*, *nakhlites*, and *chassignites* are relatively calcium-rich, and thus, interpreted to have resulted from meteorite impacts on the Moon and Mars of sufficient energy to have propelled pieces of the lunar and martian crust to orbital escape velocity. Sophisticated geochemical arguments lend support to this inference. If true, then pieces of the martian crust are found on Earth! This still leaves unanswered the question of how Earth's crust originated.

The age and composition of the crust varies with geologic setting. The crust in ocean basins is young, less than 150 million years, and dominantly basalt or *amphibolite*, a metamorphic equivalent of basalt, with a thin veneer of deep-sea sediments. Oceanic crust is also fairly thin, about 10 kilometers or less.

Continental crust is much older and thicker, and generally of granitic (silica-rich) composition. However, its lower portions are probably more mafic (something like gabbro) than are the upper portions. In most regions there is a veneer of sedimentary rocks a few kilometers thick overlying the igneous basement rocks. Because continental crust is much thicker and less dense (2.7 g/cc) than oceanic crust (3.0 g/cc), it is more buoyant and rides higher on the underlying asthenosphere than does oceanic crust, thus explaining the elevation difference between ocean basins and continents.

Several lines of evidence indicate that the crust is derived from the upper mantle. Igneous activity transforms rocks of the upper mantle into oceanic and continental crust. Evidence that this process continues currently is the widespread volcanic activity around the globe, the locations and compositions of which are consequences of plate tectonic activity.

The Mechanism of Plate Tectonics

The theory of *Plate Tectonics* has revolutionized the way in which we think about Earth. The basic mechanism (see Figure 6.4) was elucidated by Harry Hess in 1962. *Plates* of lithosphere are formed by convective upwelling of hot material from the mantle to the surface at the *mid-ocean ridges*. From this zone, plates move away at rates of a few centimeters per year, and eventually become cold and dense enough

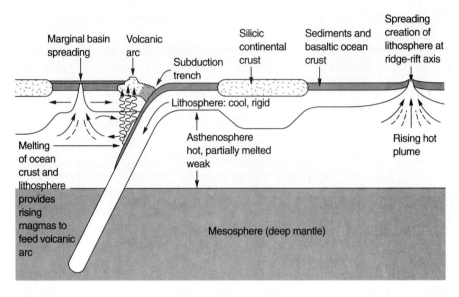

FIGURE 6.4
The mechanism of plate tectonics. The lithosphere, a plate of solid rock, is created at the spreading ridge-rift axis, and moves outward from that point riding on the partially molten asthenosphere. Less dense continental crust rides higher than oceanic crust. Eventually the oceanic lithospheric plate becomes cold and dense, subducting beneath another plate, forming an oceanic trench and volcanic arc behind the trench.
Source: Press, F., and R. Siever (1985). *Earth*. San Francisco: W.H. Freeman. Fig. 19.3, p. 461.

to subduct, or descend into the mantle, beneath another plate of lithosphere. The locations of *subduction zones* are marked by *deep ocean trenches*. The continents are passive passengers on these conveyor belts of lithosphere and ride high and buoyant because of their low density. When two plates carrying continental crust on their converging margins collide along a subduction zone, mountain belts form because the crust is shortened and thickened, and the buoyant crust of one plate is thrust beneath the other, causing uplift of a mountain belt. The locations of plate boundaries are identified by seismic and/or igneous activity. The nature of igneous activity at each type of boundary is discussed in Chapters 7 and 8. The composition of the crust is granitic (continents) and basaltic (oceans)—how likely are these compositions products of partial melting of the upper mantle?

6.5 EXPERIMENTAL STUDIES OF PARTIAL MELTING

Several laboratory experiments on the role of partial melting in the upper mantle provide a clue to its importance in crustal development. These experiments have profound implications for understanding how Earth functions, and also strengthens confidence in inferences about its interior.

Let's assume that the upper mantle at the depth of the asthenosphere is a garnet peridotite, a common rock type in ultramafic xenoliths. Laboratory experiments in which the temperature is slowly raised while pressure is maintained at the equivalent for this depth reveal that droplets of melt first appear at the junction between crystal boundaries. The transition from solid to liquid decreases the rock's density about 10 percent as a result of a 10 percent volume increase. This physical change may promote fracturing of the solid residuum, through which the lighter fluid fraction may rise and coalesce as magma pools at higher levels in the mantle or crust. Of even more profound importance is the fact that with partial melting of 5 percent to 10 percent, the composition of the liquid is basaltic. It surely is not coincidence that the most widespread rock in the crust is basalt. Other studies have demonstrated that partial melting of basalt can produce a granitic liquid. Thus, young oceanic crust can be regarded as a more "primitive," or "less refined," version of the (generally) much older continental crust.

SUMMARY

Using a combination of direct and indirect evidence, geologists have woven a tapestry illustrating Earth's structure and composition. Having constructed the picture, they have sought to test it with additional evidence. This evidence comes from diverse sources, and as we continue our exploration of our solar system, we may find additional evidence filling the gaps of the yet unfinished tapestry. The clues range from features as obvious as a volcano to seemingly unrelated bits and pieces of past and present planets fallen from the sky as meteorites.

7
Volcanism and Volcanic Rocks

7.1 INTRODUCTION

One of the most spectacular exhibitions of natural phenomena on Earth is a volcanic eruption. The eruptions of volcanoes provide many clues to the internal workings of the Earth, and the products of volcanic eruptions serve in a variety of ways as the means of deciphering Earth's history. In the previous chapter we discussed some of the products of volcanism and how fragments of the upper mantle are used to determine Earth's internal composition and structure. Beyond this, volcanism has a much broader role in contributing to the growth of the continents; to the formation of continental crust; and because of the large volume of volatiles produced during volcanic eruptions, to the formation of Earth's atmosphere and oceans. Volcanism is a primitive mechanism, producing new rock and new crust where none existed before. Unlike other rocks requiring a pre-existing rock or material which is then subsequently modified by erosion and transportation (*sedimentary rocks*), or temperature and pressure (*metamorphic rocks*), *volcanic rocks* are produced by cooling of molten material.

7.2 SCIENTIFIC IMPLICATIONS OF VOLCANIC ROCKS
Plate Tectonics

Volcanic rocks are produced as a result of lithosphere plate movement, and study of volcanic rocks enables understanding of the plate tectonic cycle. *Basaltic* volcanism is dominant at *mid-ocean ridges* where plates of lithospere form and subsequently move away in opposite directions, eventually to subduct beneath a converging plate. Above subduction zones, *composite volcanoes* (compositionally and/or texturally layered volcanoes) of *andesite lava* and ash are common. Basaltic volcanoes can also occur within plate interiors, either on oceanic crust or on a continent. The Hawaiian Islands are probably the best example of basaltic intraplate volcanism.

Radiometric Dating

Volcanic rocks containing abundant feldspar are some of the most easily radiometrically dated of rocks, providing a measure of time's passage through geologic history. The basic principle of radiometric dating results from the natural, spontaneous decay or loss of subatomic particles in elements that occur as unstable radioactive isotopes. A rock's chronometric age can be determined if the ratio of radioactive parent (original) isotope to radiogenic daughter (product) elements can be measured, and if its constant rate of decay is known. This ratio is higher for young rocks and decreases as the age of the rock increases. Volcanic and shallow intrusive igneous rocks provide one of the best ways to link together the relative *geologic time scale* (Appendix I) and the *absolute* or chronometric (measured units of time) *time scale* made possible by radioactive clocks.

History of Earth's Magnetic Field

Because most volcanic rocks have a high magnetite content, especially the very abundant basalts, volcanic rocks also provide a *geomagnetic time scale* that has been used in a number of ways to support the theory of plate tectonics. This history of the Earth's magnetic field is stored in the *remanant magnetism* of the magnetite (Fe_3O_4), acquired when the grains of magnetite cooled below the *Curie Point* (500 to 600°C). By matching the magnetic and age data derived from volcanic rocks, geologists and geophysicists have been able to construct the so-called "polar wandering curves" which depict the *apparent* polar, wandering paths of the Earth's magnetic poles relative to individual continents. However, the discovery that each continent appeared to have its "own" magnetic pole positions over the past 250 million years showed that it is the *continents* that have wandered, rather than the magnetic poles. Because the polar wandering paths for each continent are unique, comparing polar wandering paths of continents (for example, Europe and North America) indicates that the continents must have been joined together at one time, as well as the time at which they subsequently began moving apart to form the Atlantic Ocean.

Volcanic rocks are also the best tools available to date *geomagnetic reversals*—the changes in the polarity of Earth's magnetic field. The N–S polarity reversal time scale assembled from these studies in conjunction with the remanant magnetism of the ocean floor basalts enabled the determination of spreading rates of mid-ocean ridges. This can be done for any segment of time for which there are polarity reversal data (200 million years ago to the present).

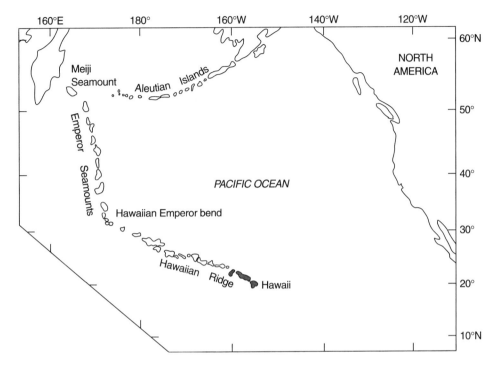

FIGURE 7.1
The Hawaiian Islands–Emperor Seamount Chain
Source: Clague, D. A., and G. B. Dalrymple. 1987. *U. S. Geol. Surv. Prof. Pap.* 1350, p. 6.

Hotspot Tracers

Volcanoes record the movement of lithospheric plates in yet another way—by tracking the *apparent* movement of hot spots. *Hot spots*, which are immense bodies of hot rock originating from fixed sites within the mantle, are areas where melts are produced as the body reaches the base of the crust to produce a huge volcanic "pimple" on the surface. The track and direction of a moving plate relative to the fixed hot spot can be easily identified by the line of volcanoes produced and the gradation of their ages from youngest at the volcanically active end of the track to oldest at the other, where long inactive volcanoes have been worn down by erosion and subsided beneath sea level as they cool after moving off of the active hot spot.

The best example of such a hot spot track is the Hawaiian Islands, in the Emperor Seamount chain (Figure 7.1). The northern-most seamount of the Emperor chain has been dated at about 80 million years of age (Clague & Dalrymple, 1981). A *seamount* is a flat-topped volcanic edifice truncated by wave erosion and now subsided below sea level. The youngest volcanic island in this chain is the large island of Hawaii where new volcanic material is produced every year. If we look beneath the waves, a younger volcano, Loihi, southeast of the Big Island will some day form a new island. This chain is also of interest because it has a bend in it in the vicinity of Midway Island. This 'elbow' records a change in the direction of plate motion approximately 40 million years ago.

7.3 CHEMISTRY AND PHYSICS OF MAGMA

Magma can be defined as naturally occurring, mobile rock material that is at least partially molten. The requirement of at least some molten rock is necessary to distinguish magma from other rocks (such as shale or bedded halite) which may be mobile forming diapirs (rising balloon-shaped masses) similar to magmatic intrusions. A *rock* is an aggregate of one or more minerals. Some rocks may appear to be *monomineralic*—composed of only one mineral—at first glance, but even with a hand lens or sometimes a microscope, only the most abundant minerals are identifiable. *Igneous rocks* form when magma cools and solidifies, or crystallizes. The volume of an igneous rock is approximately equal to the volume of magma from which it cooled.

The most abundant and widespread magma type is basaltic, containing between 45 and 55 weight-percent SiO_2. Basaltic magmas are the hottest known magmas, reaching Earth's surface at temperatures sometimes exceeding 1200°C. At the opposite end of the spectrum are the silica-rich *rhyolites* (70 to 75 weight-percent SiO_2) which have eruption temperatures of 650°C to 850°C. Closer to the source, and beneath the insulating rock of the deep crust or mantle, all of these magma types could be hotter.

The silicate content of a magma determines its viscosity. *Viscosity* is defined as a measure of the resistance to flow. The unit of measure is the *poise*. Studies of silicate melt behavior suggest that the bonds between silica and oxygen form long chains that increase viscosity. Water at 25°C has a viscosity of one poise. Asphalt on a hot summer day has a viscosity of about one million poise, which is about that of rhyolite. By contrast, hotter basaltic magma, having less silica, typically has a viscosity of 100 to 1000 poise.

The effect of water and other volatiles such as chlorine and fluorine on magma viscosity and eruptive behavior is dramatic and important. All magma contains at least some dissolved water and other volatiles. Because water is a polar molecule it is particularly effective at breaking the silicon-oxygen bonds and reducing the viscosity. The same is true of chlorine and fluorine. The physical effects of these volatiles are also important because they drive the eruption of magma by expansion as it rises, and thus, as the pressure on it declines. In some cases the expansion can be explosive, causing the magma's disaggregation into small clots when ejected from the vent. Cooling as it travels through the atmosphere, it accumulates around the vent as *cinders*. This behavior can occur with any magma type but the tendency for explosive behavior increases with increasing silica content. Thus, andesites (55 to 62 weight-percent SiO_2) are more explosive than basalts, and rhyolites are more explosive than andesites.

In addition to basalt, *rhyolite, dacite* (62 to 70 weight-percent SiO_2) and *andesite* are common magma types. Some other silicate and nonsilicate magmas are less common. These include the *alkalic* (rich in Na and K) and *peralkalic* (rich in Na and K, poor in Al_2O_3) magmas, also silicate magmas, but less common than the basalt rhyolite *calcalkalic* series. Other, nonsilicate magmas include sulfur magmas and magnetite-apatite magmas. The latter are believed to form by immiscibility (nonmixing) of an iron-rich liquid with silicate liquids. Evidence for this immiscibility hypothesis comes from laboratory experiments and the formation of glassy, iron-rich globules in basalts. The rich iron deposits produced by this magma fraction are a source of iron ore [e.g., Iron Mountain of the St. Francois Mountains (Ozark Mtns.), Missouri].

7.4 TYPES OF VOLCANOES AND VENT AREAS

The material ejected from volcanic vents takes several forms including lava, blocks and bombs, ash, pumice, and gases. With the exceptions of the gases, most of these materials accumulate in the vent's vicinity, forming conical or near-conical structures. In some instances, very large volumes of magma are ejected in short periods of time from a shallow magma chamber, and thus, produce circular or subcircular collapse zones or *collapse calderas* (e.g., Yellowstone Caldera).

Shield Volcanoes

The largest volcanic edifices on Earth are *shield volcanoes*. These structures have the domal profile of a medieval warrior's shield (Figure 7.2) and are composed mostly of solidified low-viscosity lavas. Most commonly this lava is a basalt. Typically, the slopes of a shield volcano range from 5 to 10 degrees. On Earth, the best known and largest examples are the Hawaiian Islands. Rising from the ocean floor 15,000 feet deep, and composed of basaltic rock, they attain elevations of almost 14,000 feet (total relief of nearly 29,000 ft). The portion of the volcano visible above sea level is less than half the height of the cone, the much more voluminous bases hidden beneath the waters of the Pacific Ocean. For example, the island of Oahu on which the city of Honolulu is located is composed of two large shield volcanoes, Waianae and Koolau. These volcanoes originated as two separated cones coalescing into one island while growing to their present size. The large island of Hawaii is composed of five overlapping, coalescing cones; the two largest structures are Mauna Kea, and Mauna Loa (Lockwood & Lipman, 1987; Figure 7.3). The summit of each cone is capped by a summit caldera or summit crater which, from time to time, harbors an active lava lake. Commonly, eruptions on Hawaiian volcanoes take place on the flanks. These *flank eruptions* are located along *fissures* on the cone sides that allow extrusion of the rising magma. Small "parasitic" cones or craters form on the flanks; one of the most active is Kilauea Iki, on the southeast flank of Kilauea.

Historic eruptions in the Hawaiian chain have occurred only at Kilauea and Mauna Loa, except for a brief eruption of Hualalai Volcano, also on the Big Island,

FIGURE 7.2
The Sierra Grande shield volcano composed of fluid andesite, in Raton–Clayton Volcanic Field, northeastern New Mexico. The volcano is approximately 10 km in diameter and rises almost 650 m above surrounding volcanic plain.

in 1800–1801. Since 1950, Hawaiian volcanic activity has occurred only at Kilauea, except for two brief eruptions of Mauna Loa in 1975 and 1984. The activity at Kilauea has been nearly continuous since 1820, when the first American missionaries established missions in Hawaii and reliable records were begun. The only period of sustained quiescence since then was between 1934 and 1952. However, Mauna Loa was active during that period, but dormancy began in 1952, lasting until 1974 when a minor eruption was recorded (Lockwood & Lipman, 1987).

Monitoring and Predicting Hawaiian Eruptions The Hawaiian Volcano Observatory was established in 1912 to monitor and study Hawaiian volcanism. Operated and managed by a number of government agencies and bureaus during its first 36 years, it has been under the guidance of the United States Geological Survey, a branch of the Department of the Interior, since 1948.

Monitoring and predicting eruption involves a number of techniques including visual monitoring at the vent; recording and analyzing seismic activity; ground deformation measurement; and geochemical methods involving monitoring the change in vent gas volume and composition (Tilling, 1987).

Volcanism and Volcanic Rocks

Ground deformation, or changes in the shape of the volcano, are determined by *tiltmeters* located on its flanks. The units of measure are microradians (equal to 0.00006 degree). Tilt changes occur as magma flows into the volcano and it "inflates." Tilting is a medium- to long-term process that may last for years, but more commonly occurs within a matter of weeks prior to eruption. Once magma begins to flow on the surface, the volcano begins to deflate (Figure 7.4). Deflation

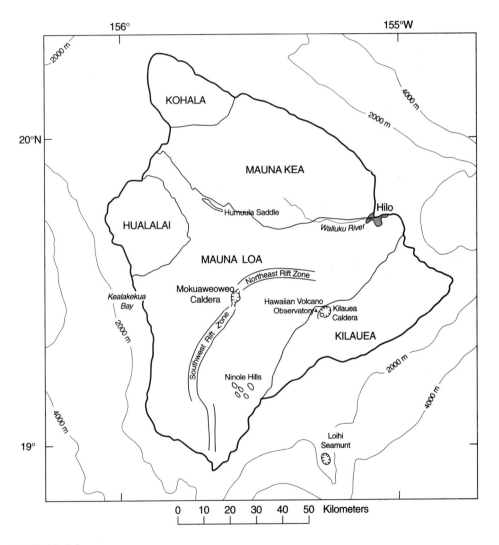

FIGURE 7.3
Map of the large island of Hawaii showing the five volcanoes comprising it. Historic volcanism restricted to large island.
Source: Lockwood, J. P., and P. W. Lipman. 1987. *U. S. Geol. Surv. Prof. Pap.* 1350, p. 517.

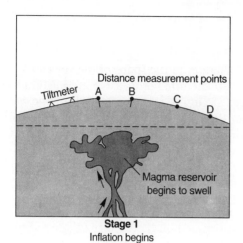
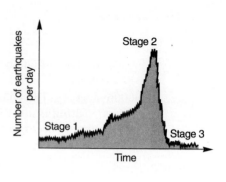
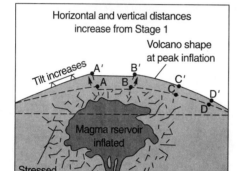
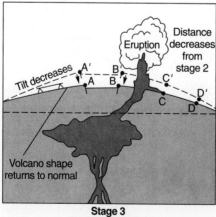

FIGURE 7.4
Diagram showing inflation and deflation of volcano and its influence on tiltmeters and laser ranging instruments.
Source: Tilling, R. I. 1987. *Monitoring Active Volcanoes*. U. S. Geological Survey General Interest Publication.

occurs much more rapidly than inflation. Another indication that the volcano is changing shape as magma moves upward into the cone is the changing diameter of the caldera rim. This distance is monitored by laser beams bounced off a reflector on the opposite rim. The period of the beam's travel time is measured and converted to distance with an accuracy of 1 cm per 10 km.

The most useful technique for monitoring volcanic activity of all types is the *seismograph*. This instrument continuously monitors and records seismic activity and produces a graph that geophysicists interpret. Information collected from a network of seismographs includes time and number of earthquakes as well as their depths and locations. Seismic analysis allows monitoring of an eruptive episode from the beginning of the magma movement from the upper mantle to the surface. Magma movement produces a distinctive seismogram trace distinguishable from microearthquakes or longer-period earthquakes of nonvolcanic origin. Portable seismometers positioned on the volcano's flank prior to an eruption allow geophysicists to predict the location of magma extrusion.

The Mauna Ulu Eruption of Kilauea Volcano: 1972–74

The Mauna Ulu eruption episode, one of numerous eruptive periods in recorded history, provides insight into Hawaiian volcanic phenomena. It would be inaccurate to call this or any other episode "typical," because of the wide variety of possibilities anytime an eruption occurs. However, most eruptions have several common features such as lava fountains, development of lava tubes and channels, and the flank eruptions and eventual flow of lava into the sea typical of Hawaiian eruptions.

The following discussion is derived from the treatment by Tilling et al. (1987).

The Mauna Ulu eruption began in February, 1972, after several months of inflation and seismic activity beneath the eastern rift (Figures 7.5 and 7.6). Lava extruded at Mauna Ulu and then at Alae Crater. These two craters were apparently connected by a lava tube, lava from one crater feeding the other. Lava overflowed the craters several times and traveled a short distance before draining back into the vent.

Soon a lava lake formed in each of the craters fed by fountains of lava, and on March 18, 1972, a new fissure vent more than 600 ft long opened on the west side of Mauna Ulu, producing fountain activity. Eruption of lava, however, soon shifted to Alae Crater. This change was predicted by summit deflation and harmonic tremors in the Alae Area. Harmonic tremor gives a seismograph signature indicating magma movement beneath the surface. The lava lake began to fill and overflow, followed by flow back into the crater. Each of these overflow-drainback episodes lasted from 30 to 120 minutes.

In June, 1972 a new phase began. The lava lake level at Mauna Ulu dropped and a system of lava tubes and channels developed, draining lava from Alae Crater into Makaopuhi Crater. By early August the flow into Makaopuhi had declined and the lava moved to the southeast toward the shoreline. By mid-August the flow had crossed Chain of Craters Road, a point halfway to the coast, and on August 23 it flowed into the sea at Kaena Point. This phase continued through October, when the flow volume decreased and the bulk of lava flowed away from the tube system. On October 20 lava flowing through this system of tubes and channels ceased flowing into the sea. Where would it go next?

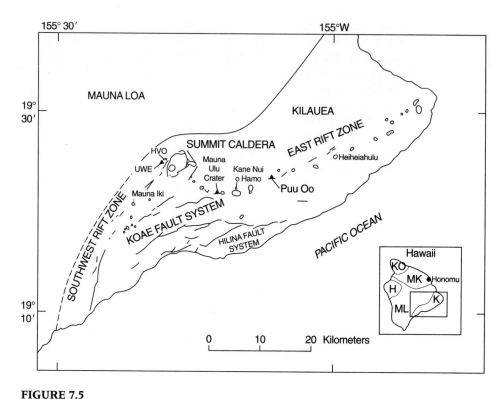

FIGURE 7.5
Map of the Kilauea Volcano
Source: Tilling et al. 1987. *U. S. Geol. Surv. Prof. Pap.* 1350, p. 406.

For the next couple of months, lava from Alae Crater flowed into Makaopuhi crater, and also built up the Alae volcano with multiple small flows. Toward the end of this phase a new lava tube and channel system developed to the south, and on February 24, 1973, lava again reached the sea—this time near Apua Point. This activity continued until May 1. A strong earthquake (magnitude 6.2) on April 26 probably disrupted the lava's flow path, shutting off flow to the sea a few days later. Tube blockage occurred just above Poliokeawe Pali causing overflows upstream from this point through holes in lava tube roofs (*skylights*). The overflows traveled a short distance eastward cascading down the side of Poliokeawe Pali and Holei Pali. A brief lull from May 3 until the morning of May 5 ended when lava erupted from Pauahi Crater after a series of small, shallow earthquakes. Lava fountains along an eastern fissure reached a height of 20 m. By early afternoon this activity died down and lava fountains up to 50 meters high played along a set of fissures near Hiiaka Crater (Figure 7.6).

The eruption episode continued with many more shifts of vent areas and other twists and turns until July, 1974. The activity was accompanied by repeated filling and draining of lava lakes at Mauna Ulu and Alae, fountaining episodes, and crater overflows.

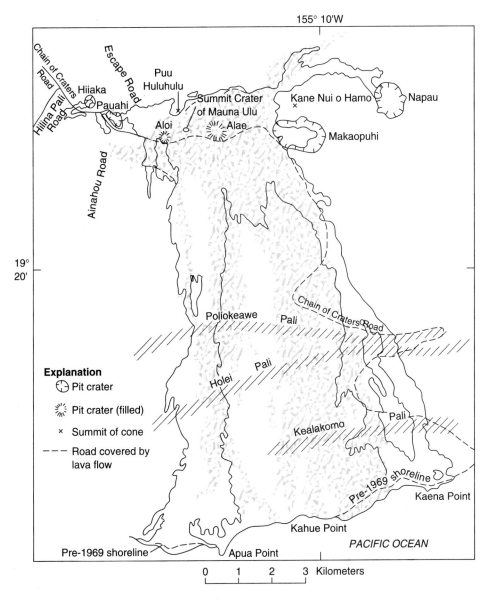

FIGURE 7.6
Map of the Mauna Ulu and related vents on southeast flank of Kilauea, showing extent of lava flows from 1972–74 activity (outlined by solid line). Shaded areas show lava fields produced by 1969–71 Mauna Ulu eruption.
Source: Tilling et al. 1987. *U. S. Geol. Surv. Prof. Pap.* 1350, p. 407. Modified from Swanson et al. (1979) and Holcomb, R. T. (1976).

FIGURE 7.7
Volcano Fuego on outskirts of Antigua, Guatemala, showing classic profile of composite volcano.

 The eruption was significant in its well integrated lava tubes and channels, probably the result of sustained flows allowing sufficient time and lava volume to produce a continuous lava channel. The episode did lack the high fountains (more than 500 m high), a part of many Hawaiian eruptions. Little net summit inflation indicated an almost constant influx of magma into the system. Whenever magma flow shifted to another part of the system, a local area inflated much like a small blister beneath a thin layer of skin. An almost constant lava supply is also suggested by the sequence of events—whenever one vent shut down or its activity declined, another would start up or increase its activity.

Composite Volcanoes

The *composite volcano* is composed of two types of materials, ash and lava. These are the typical structures associated with the generic term, volcano, most of which form above active subduction zones. Examples of composite volcanoes include Mt. Fuji in Japan, Mt. Etna in Sicily, Mt. Rainier and Mt. St. Helens in the Cascade Mountains of Washington, and Mt. Vesuvius in Italy (also see Figure 7.7). Though not as high or voluminous as shield volcanoes, they may attain heights of 20,000 feet. Their slopes are much steeper than those of shield volcanoes, ranging between 25 and 30 degrees. Lava and ash are composed of andesite or dacite, a more silicic, viscous, and explosive magma type. The magma typically reaches the surface at temperatures of 800 to 1000°C.

Mount St. Helens   Mount St. Helens is one of the most intensively studied volcanoes because of its location and the active cycle that began in 1980 (Figures 7.8 and 7.9). The eruptive history of one of the most active volcanoes in the 48 contiguous states is useful for the information it provides about composite volcanoes and how they differ from shield volcanoes. Proximity to densely populated areas and its accessibility generated cause for concern when the first earthquake and minor eruption were recorded in early 1980.

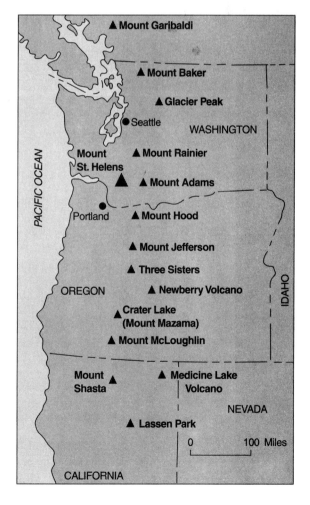

FIGURE 7.8
Map of the Cascade Range showing regions of major volcanic peaks.
Source: Tilling, R. I. 1987. *Eruptions of Mount St. Helens*. U. S. Geol. Surv.

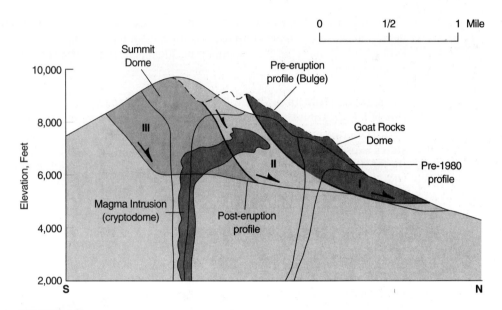

FIGURE 7.9
North–South profile of Mt. St. Helens showing bulge at maximum extent and pre-1980 profile (dashed). Block labeled I was first to slide, followed by blocks II and III.
Source: Tilling, R. I., 1981. U. S. Geol. Surv.

Mt. St. Helens' history can be determined by studying its lava flows and pyroclastic material of which it is comprised. The following history, from the studies of Donal Mullineaux and Dwight Crandell (1981), reveals the story of a very active volcano about 40,000 years old—considerably younger than others in the Cascade Range.

> Dacitic volcanism, consisting mostly of pyroclastic flows, dominated the first 35,000 to 37,000 years of its history. Although there is evidence of a dormant interval from about 35,000 to 20,000 years ago, the lack of volcanic deposits from this period may reflect erosion rather than nondeposition. The stratigraphic sequence for the last 10,000 to 13,000 years is more complete, revealing several dormant intervals of 200 to 800 years' duration. The dormant intervals are separated by periods of activity including dacite, andesite, and some basalt lava flows; dacite domes; and pyroclastic flows of dacite and andesite. The last eruptive episode prior to 1980 is the Goat Rocks Eruptive Period which began about 1800 A.D. This episode is characterized by a dacitic pumice layer found as far east as northern Idaho. These pumice eruptions were noted by trappers and explorers in the region between about 1800 and 1857—the last eruption dating prior to 1980.
>
> The events leading to 1980 had produced, during the span of approximately 40,000 years, a beautifully symmetrical volcanic cone with a relief of more than 5,000 feet. The volcano had been dormant for 123 years when, on March 20, 1980, a magnitude 4 earthquake beneath it was recorded.

The chronology of events in the first half of 1980 is condensed from the narrative by Christiansen and Peterson (1981):

> An earthquake swarm quickly followed and on March 25, 24 earthquakes of magnitude 4 or greater were recorded in 8 hours. This rapidly awakening volcano first erupted on March 27, producing a column of volcanic ash rising more than 6,000 feet. The summit was not visible then because of dense clouds, but when they cleared, aerial reconnaissance revealed a new crater near the snow-covered summit. An east-west fracture more than 4,000 feet long bisecting the summit area was also visible. Another explosive ash eruption occurred the next day. Prevailing winds from the west blew most of the pyroclastic material onto its east flank, making the summit snow-white on the west half and darkened by blanketing ash on the eastern half. By late March a second crater had formed near the summit, and shortly thereafter both enlarged, forming one large crater more than 1,000 feet in diameter.

Also in late March a curious phenomenon that was to gain prominence in the next two months first became evident—a bulging of the volcano's north flank. This bulge was first noted on March 27, and it continued growing at a nearly constant rate of five to six feet per day. The bulge's growth apparently resulted from magma moving into the north flank beneath an older dacite dome, and possibly filling an older summit crater. By May 12 the Goat Rocks, an old dome on the north flank, had been moved more than 330 feet northward by the growing bulge. Many small ash eruptions, accompanied by earthquake activity with harmonic tremor, continued intermittently until the climactic May 18 eruption. Many of the smaller eruptions consisted of a white steam cloud, an upper dark ash cloud, and a ground-hugging turbulent ash flow that traveled rapidly down slope.

The Climactic Eruption of May 18, 1980   May 18 was a bright clear spring morning in the Pacific Northwest. The volcano showed no evidence of impending eruption given the values of the multiple parameters being monitored—seismicity, deformation, and gas chemistry. About 8:30 a.m., a magnitude 5.1 earthquake was recorded about one mile beneath the volcano. About 20 seconds later the bulge began collapsing in a series of steps, releasing the pressure within the magma beneath it and triggering one of the most explosive, spectacular, and devastating eruptions in recorded United States history.

A photographic sequence of the summit reveals that the entire north flank collapsed in a series of blocks (refer again to Figure 7.9). Resulting from this collapse, the pressure on the magma doming beneath the bulge decreased, the resulting effect similar to popping the top of a warm, shaken container of carbonated soda. The gas bubbles expanded explosively in response to this suddenly decreased pressure, feeding the magma's eruption. The first magma erupted from the vicinity of the fault planes along which the collapse took place. This collapse produced a large avalanche that moved northward at more than 150 miles per hour toward Spirit Lake, across it, and into the upper reaches of the North Fork of Toutle River. On the heels of this avalanche was the lateral volcanic blast, directed northward and slightly upward, and propelled by rapidly expanding gases. This volcanic component consisted of hot

FIGURE 7.10
Mt. St. Helens in eruption on May 18, 1980
Source: Courtesy of the U. S. Geological Survey. Photograph by Krimmel.

gases, ash, and rock fragments torn from the volcano's flank. The eruption affected an area to the north as far as 20 miles away from the summit. The velocity of this lateral blast initially exceeded 600 miles per hour, but subsided quickly from wind resistance.

Following this northward, vent-clearing blast, the volcano erupted vertically, spewing ash quickly to heights exceeding 12 miles (Figure 7.10). At various levels where the ash plume encountered changes in atmospheric density, it began spreading into a mushroom shape, a typical *plinian* eruption column. Prevailing westerly winds carried the cloud eastward and then southeast, dropping ash particles. This ash plume's eastward growth could be traced on satellite images as far as eastern Montana. The ash deposited by this plume was detected on clean, smooth surfaces such as car hoods and roofs as far away as Oklahoma and Texas a few days after the eruption. Strong explosions fed the column of ash throughout the day but subsided after several hours. The ash volume erupted about 0.3 cubic mile, less than 10 percent of the landslide's volume. By the morning of May 19, one of the most spectacular and most highly monitored of volcanic eruptions had ended. Mt. St. Helens' summit elevation on May 19 was about 8,364 feet, or 1,313 feet lower than it had been 24 hours earlier. The new crater blasted out to the north measured two square miles across and about 2,100 feet deep. The terrain to the north more than six miles was entirely devoid of vegetation, covered by a thick layer of ash and rock deposited by the volcanic flow that followed the landslide. The surface was pocked by numerous small craters from steam explosions caused by red hot ash contacting pockets of water. These steam explosions continued for weeks after the eruption. Measurements two weeks after the eruption revealed that these ash deposit temperatures ranged from 290°C to 420°C.

Shortly after the catastrophic eruption, a dome of viscous lava began building in the large crater it had produced. This dome growth has lasted several years.

The volume of material erupted from Mt. St. Helens during the May 18 eruption was fairly small compared to eruptions of other volcanoes. For example, the 1815 eruption of Krakatoa in Indonesia produced almost seven cubic miles of ash and pyroclastic debris, much of which circled the world several times, producing slightly cooler temperatures and redder sunsets for years. However, one of the largest eruptions of composite volcanoes was from prehistoric Mt. Mazama 7,000 years ago, an event ejecting more than 10 cubic miles of ash. This episode partly destroyed the upper part of its cone, the remaining portion collapsing into the evacuated magma chamber, producing what is today known as Crater Lake (Figure 7.11). This is the

FIGURE 7.11
View northwest of west wall Crater Lake, southern Oregon, produced by the eruption of Mount Mazama 7,000 years ago.

deepest lake in North America, reaching a maximum of 1,900 feet. Although the amount of magma produced by individual eruptions of composite volcanoes may be less than that of other types, the pyroclastic eruptions from a composite volcano are one of the most spectacular and deadly of all natural phenomena.

Cinder Cones

Cinder cones are small (500 to 3,000 ft) structures composed of cinders. *Cinders* form from small clots of magma of sand to pebble size, thrown out of the vent and rapidly cooled while falling through the atmosphere. *Cinder cones* are cinder accumulations around a vent. The cones form slopes at the angle of repose, approximately 25 to 35 degrees. Most cinder cones are one-event structures, forming during one interval of volcanism. Unlike composite volcanoes that may remain dormant for hundreds or even thousands of years, cinder cones generally do not reactivate after falling quiet. Many cinder cones have associated lava flows. In some cases, multiple lava flows may have breached one side of the cone, or more commonly, may have flowed from the cone's base. For example, Mt. Capulin in northeastern New Mexico has three distinct lava flows that originated from the cone's western base between 5,000 and 10,000 years ago (Figure 7.12).

Numerous examples of cinder cones occur throughout western North America. Most of these structures are less than one million years old. Probably the most famous of all cinder cones is Paricutin, one of the few volcanoes having erupted in historic times where none existed before.

FIGURE 7.12
View North of Mt. Capulin, cinder cone 1,000 feet high, northeastern New Mexico. Note west side of crater is lower than east side, probably the result of a west wind during an eruptive episode.

The Birth and Growth of Paricutin Volcano Extending across Mexico, in an east-west line at Mexico City's latitude, is a line of active volcanoes whose history extends back to the Late Tertiary, from about 2 to 25 million years ago. This so-called *Trans-Mexican Volcanic Belt* includes 11 volcanoes that had been historically active through 1942. These volcanoes are the result of subduction of the Cocos and Rivera plates beneath the North American Plate off the southwest coast of Mexico.

In 1943 a new volcano, Paricutin, was born in the western part of the belt in the state of Michoacan. The following narrative is adapted from a discussion by Foshag and Gonzales (1956):

> Earthquakes began the first week of February, continuing for a couple of weeks. On the afternoon of Saturday, February 20, a fissure opened, emitting sulfurous gases and steam, followed shortly thereafter by a column of volcanic ash and blocks, probably from the fissure's walls. The vent gradually grew larger, as did the cone of accumulating material around the vent. Lava first reached the surface sometime during the night of February 20. At midnight the cone was almost 20 feet high, and by 8 a.m. the next morning it was an estimated 30 to 36 feet high. The volcano, like a growing young boy, went through several explosive periods, before attaining a mature volcanic structure's shape. By the afternoon of the second day the cone was more than 150 feet high and the material ejected from the vent changed from ash to volcanic bombs tossed to heights of more than 1,500 feet.

Volcanic bombs are clots of magma, larger than a cinder, which are ejected from a volcano and acquire a generally rounded shape in their flight through the atmosphere. Some accounts of the growing volcano on February 22 report two vents within the crater: a larger vent producing pyroclastic ejecta, and a smaller one at the northeast base of the cone from which lava flowed. By February 26 the cone measured more than 500 ft high and more than 2,000 ft basal diameter. Lava and ash continued ejecting from the new volcano so that by the end of its first year, lava flows reached as far as three miles to the north and the crater had attained a height of almost 1,200 feet. The village of Paricutin was evacuated in April 1943, and the village of San Juan Parangaricutiro, three miles to the cone's north, was evacuated in the face of advancing lava flows in June 1944. By March 4, 1952, when volcanic activity ceased, the cone had grown to a height of 1,300 feet.

Collapse Calderas

Unlike the positive topographic features of the volcanoes, collapse calderas form broad areas where the surface has collapsed. This topographic difference is commonly masked or obscured because younger deposits have filled in the *caldera*—a large oval or circular depression bounded by faults.

The large collapse calderas associated with rhyolitic volcanism are not to be confused with the small summit craters or calderas that crown all volcano tops, from cinder cones to shield volcanoes. Although these might also be called calderas, these small collapse zones of volcano summits are an order of magnitude less in size than are collapse calderas. More properly they should be referred to as summit calderas or summit craters, thus preventing confusion with the 10- to 50-mile diameter collapse calderas associated with large ash-flow tuff eruptions.

The magma associated with these large structures' formation is *rhyolite*, a magma type rich in silica and sodium or potassium. The magma erupts from fissures in large volumes as ash or pyroclastic flows. An *ash flow* is a mixture of ash, gases, pumice, and rock fragments similar to that created by composite volcanoes. However, the volume of the eruptions is much greater, evacuating the shallow magma chamber with such rapidity and efficiency that its roof collapses.

The sequence of events is illustrated in Figure 7.13. The first step (I) is a broad regional doming as magma moves close to within a mile or two of the surface. This shallow reservoir of magma heats the country rock, causing expansion and fracturing. Step two (II) is a voluminous eruption(s) of magma as ash flows. The volume of material erupted commonly exceeds 100 cubic miles, resulting in a blanket of ash and rock that may extend more than 60 miles from the caldera and cover an area of more than 3,000 sq mi. The next step (III) in the caldera's development occurs in response to the voluminous eruption—collapse of the magma chamber roof, producing a circular or semicircular depression 1,000 feet or more deep. The area's collapse may occur in several discrete steps during eruption, or in one episode following eruption. If collapse occurs during the eruption, ash flows will tend to pond in the new depression. Regardless, the next step (IV) is the partial or complete filling of the new depression. The fill is heterogeneous, consisting of large blocks collapsed off the caldera walls (*megabreccia*), sedimentary rocks, or volcanic rocks accumulated on the caldera floor after its formation. This late volcanic activity is not unexpected because magma remains beneath the caldera and its floor is usually pervasively fractured. These fractures provide the avenues for magma rising to the caldera's surface.

The next step (V) in the caldera's evolution is resurgent doming—a step which does not occur in all calderas. It is not understood why this doming occurs or why it occurs in some, but not all, calderas. This dome is usually centrally located within the caldera. Resurgence produces a distinctive topography consisting of a generally central resurgent dome, surrounded by a moat. The last phase of activity (VI) is ring-fracture volcanism, hot springs, and SO_2 emissions that may persist for more than a million years after the caldera-forming ash flow eruption.

Examples of calderas in the western United States include the Yellowstone Caldera in northwest Wyoming, the Valles Caldera just west of Los Alamos in northern New Mexico, the Creede Caldera in the San Juan Mountains of Colorado (Figure 3.2), and the Long Valley Caldera in eastern California.

7.5 ASH FLOWS AND ASH-FLOW TUFFS

The elucidation of ash-flow eruption mechanics puzzled geologists for many years. In the western United States, large, thin sheets of rhyolitic material were noted, but observations at volcanoes showed that rhyolitic magma is extremely viscous. Rhyolitic magma form domes around the vent rather than the seemingly fluid behavior implied by individual sheets of volcanic material commonly more than 60 to 80

FIGURE 7.13

Collapse caldera sequence of formation. I) regional doming; II) voluminous ash flow eruption; III) formation of caldera; IV) pre-resurgence volcanism and sedimentation; V) resurgent doming; and VI) ring-fracture and hot springs activity.

Source: Smith, R. L., and R. A. Bailey. 1968. *Resurgent Cauldrons*. Boulder: *Geological Society of America, Inc. Memoir* 116, 1968, Fig. 5, p.634–35.

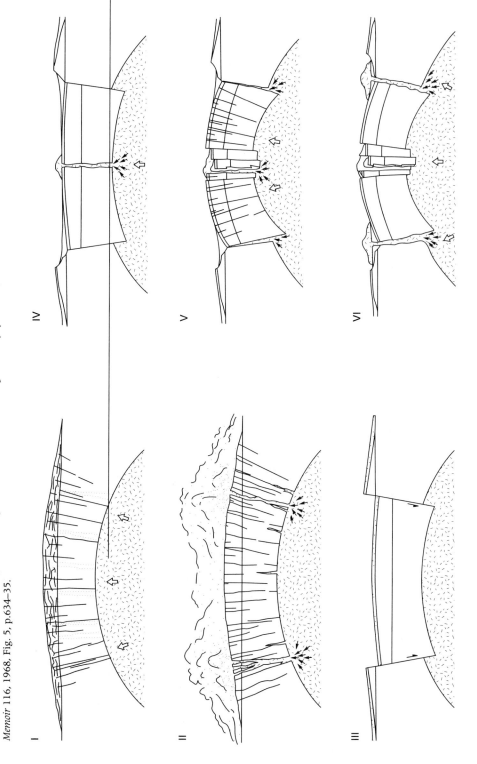

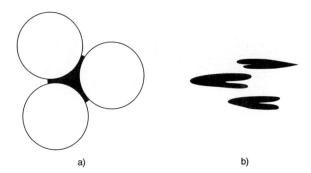

FIGURE 7.14
Morphology of uncompacted and compacted glass shards: a) uncompacted glass shards produced by breakage along thinnest part of bubble wall; b) compacted, welded glass shards.

miles wide that rarely exceed more than a few hundred feet in thickness. Thin sections of these rocks revealed a welded texture of compacted and welded glass shards (Figure 7.14). Thus, these rocks are called *welded tuffs* or *ignimbrites*. Another characteristic feature of ignimbrites is a structure called *eutaxitic texture* found in the lower portions of ash-flow sheets and consisting of flattened pumice fragments (Figure 7.15). This texture is the single best field characteristic of ash-flow tuffs. Ash-flow tuffs also commonly exhibit well-developed columnar jointing produced by contraction during cooling (Figure 7.16).

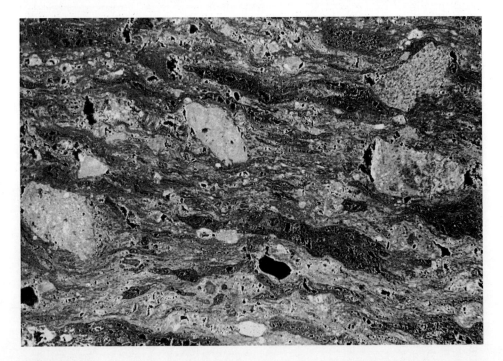

FIGURE 7.15
Eutaxitic texture, the most reliable characteristic of welded tuffs. Dark streaks are flattened pumice fragments. Large angular bodies are rock fragments. Width of photograph, 8 cm.

FIGURE 7.16
Columnar jointing in tertiary ash-flow tuff in San Juan Mountains, Colorado. Height of cliff face, 50 m.

The mechanics of these rocks' eruption were first deciphered at Mt. Pelee on the island of Martinique in the East Indies. During eruptions beginning in May 1902, small eruptions consisting of hot ash and gases moved down the slope of the volcano at high speeds (60 to 70 mph) in a turbulent manner, much like a snow avalanche. The velocity of these ash flows was enhanced by a layer of air trapped beneath the flows, enabling them to travel short distances across bodies of water! The temperature of these flows was estimated from the deformation of glass objects to be about 600°C. The resulting deposit was a layer of unconsolidated ash a few inches thick. Although the flow mechanics explained the geometry and distribution of large rhyolitic ash flows erupted in prehistoric times, the volume of the andesitic ash flows was insufficient to produce any welding or eutaxitic texture. Welding of glass shards or flattening of pumice fragments producing a eutaxitic texture have been found only in prehistoric ash-flow tuffs.

7.6 STRUCTURES AND TEXTURES OF LAVA FLOWS

The most widely recognized product of volcanoes is lava and the lava flows covering large portions of Earth's crust. The most common lava type is basaltic. In areas such as the Deccan Plateau of India or the Columbia Plateau of the northwestern United

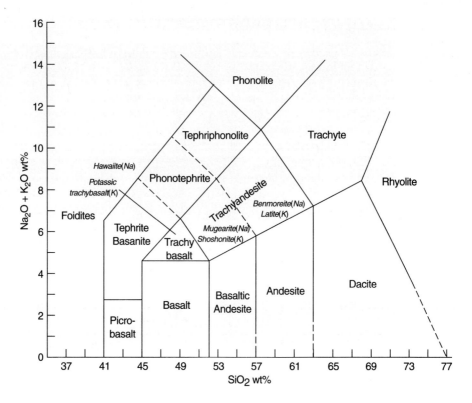

FIGURE 7.17
The total alkali-silica diagram, a chemical scheme for classification of volcanic rocks.
Source: Le Bas, et al. 1986. Fig. 2, p. 747.

States, multiple basaltic lava flows of rather uniform composition cover areas of more than 5,000 sq mi, and have an aggregate thickness of more than 1,000 feet. These flows almost invariably result from lava flowing from a fissure or set of fissures.

The surface and internal textures of basaltic lava flows also show some variation resulting from differences in temperature, gas content, and distance from the vent. The most fluid and faster moving lava flows generally exhibit a rope-like surface texture known by the Hawaiian name, *pahoehoe*. Slower moving flows, generally far from the vent, commonly exhibit a blocky texture called *aa*. When lava flows erupt underwater or flow into water, lava pillows are commonly formed through a process called *bulbous budding*. *Pillow lavas* form when the skin of the tongue of lava is rapidly chilled by the water and solidifies. However, lava in the tongue's interior is still fluid, and as the pressure builds within the bulb as fresh lava enters, a fracture forms, and a fresh tongue squirts out until the chilling water again produces a thin skin. The process repeats again and again producing a pile of many bulbous forms.

7.7 CLASSIFICATION OF VOLCANIC ROCKS

Volcanic rock classification is often considered by students as something akin to sorcery and mental torture, but the classification basis is relatively simple. With a little practice and guidance, anyone with normal powers of observation and visual acuity, and with a little help from a hand lens or magnifying glass can classify a variety of volcanic rocks. The basis of hand-specimen identification and classification for all rock types is texture and mineral composition.

The major stumbling block to volcanic rock classification is the fine-grained, sometimes glassy, texture of volcanic rocks, making it difficult or even impossible to determine their mineralogical composition. In an attempt to standardize volcanic rock classification, some researchers have proposed chemical classifications (Irvine & Baragar, 1971; LeMaitre, 1984). An example of one such scheme is shown in Figure 7.17. However, it is often desirable to classify a rock in the field; so see the Vest Pocket Guide to Volcanic Rock Classification (Table 7.1). The basis of this system is

TABLE 7.1.
The Vest Pocket Guide to Volcanic Rock Classification

Using Phenocryst Mineralogy		
Phenocryst Minerals	Typical Color	Possible Rock Types
Quartz only	Brown, Pink, Red	Rhyolite
Quartz and Alkali Feldspar	Brown, Pink, Red	Rhyolite, Qtz Trachyte
Alkali Feldspar only	Brown, Black	Trachyte
Quartz and Plagioclase Feldspar	Gray	Dacite
Plagioclase Feldspar and Alkali Feldspar	Brown	Latite
Quartz, Plagioclase Feldspar and Alkali Feldspar	Red, Brown, Pink	Rhyodacite or Rhyolite
Plagioclase Feldspar only	Gray, Purple or Black	Andesite or Basalt

Other	
Vitrophyre	Volcanic glass; usually black, with phenocrysts.
Obsidian	Volcanic glass; usually black, with conchoidal fracture.
Perlite	Volcanic glass with fractures, usually gray or brown.
Pumice	Light-colored, vesicular (very low density) rock
Scoria	Dark, usually basaltic, vesicular volcanic rock.
Tuff	Light-colored, fine-grained volcaniclastic rock composed of glass shards, crystal and rock fragments.
Welded Tuff	Characterized by welded glass shards and eutaxitic texture

FIGURE 7.18
Phenocrysts of light-colored plagioclase in fine-grained groundmass of darker minerals: a porphyritic texture. Largest phenocrysts, 3 cm long.

FIGURE 7.19
Snowflake obsidian produced by devitrification of volcanic glass. Snowflakes are a combination of silica and feldspar minerals, 4 to 6 mm in diameter.

utilization of phenocryst mineralogy to infer the rock's overall composition. The classification assumes that the *phenocrysts* (mineral grains larger than the rest; Figure 7.18) are representative of the entire rock mass. For example, a rock with quartz and alkali feldspar phenocrysts is most likely rhyolite. Color can be used as a secondary guide, but use of rock color as the only classification tool cannot be recommended.

Some volcanic rocks contain no phenocrysts but have characteristic textures that can be used to identify the rock type. These rocks are listed as "Others" in Table 7.1. In many instances a rock type and textural name can be combined to further refine the classification; e.g., rhyodacite vitrophyre or rhyolite obsidian. Most obsidian, although dark-colored, is of rhyolitic composition (Figure 7.19).

Although the philosophical basis for classification is rather simple, implementation of this classification scheme can be daunting. Even when looking at phenocryst minerals, the grains may be only a millimeter or two across. How can the minerals of such small crystals be identified? Fortunately the number of *essential* minerals is small: quartz, alkali feldspar, plagioclase feldspar, and feldspathoids. Of feldspathoids the most common is *nepheline*, which closely resembles quartz but has a greasy luster. Quartz is easily distinguished from feldspar by its conchoidal fracture. Feldspars are characterized by their cleavage, and plagioclase feldspar is distinguished from alkali feldspar by the former's twinning striations. Only practice can make one proficient at finding and confidently recognizing twinning striations. How-

FIGURE 7.20
Scoria, a highly vesicular form of basaltic lava.

ever, once the needed mental processes are mastered, the obstacles to confident identification in hand-specimens vanishes.

There is one other clue to the composition of feldspar in volcanic rock: their accessory minerals. *Accessory minerals* are those associated with the principal minerals, usually in minor amounts. The most common accessory minerals are biotite, magnetite and hornblende. The presence of hornblende, because it is a calcium-bearing silicate, suggests the presence of plagioclase. The presence of biotite, a potassium-bearing mica, suggests the presence of alkali feldspar. Proficiency in igneous rock identification, like any worthwhile enterprise, comes with practice.

8
Granites and Other Plutonic Rocks

Unlike volcanic rocks which, although not completely understood, are one of the most visible of Earth phenomena, *plutonic* igneous rocks, also called *intrusive* igneous rocks, crystallize deep underground out of our view. The interpretations that we make about their origins are thus derived from clues gathered long after these magma bodies intruded, crystallized, and were uplifted and eroded. An additional degree of uncertainty enters in estimating the amount of erosion that any intrusive body has undergone.

8.1 GEOMETRY OF PLUTONIC BODIES

Intrusions of magma, which generally occur in an upward direction because of the density contrast between the solid *country rock* (surrounding intruded rock) and the less dense intruding magma, occur in a wide range of volumes and geometries. Individual intrusions range from small *dikes* (tabular cross-cutting bodies) a few inches thick, to large masses that may cover areas greater than 100 sq mi. The general term for an intrusive body is pluton. A *pluton* represents one pulse of magma intrusion into the Earth's crust. Although generally assumed to be cylindrical in shape (Figure 8.1), plutons may assume almost any shape depending on the structure of the rocks intruded and the mode of intrusion. The terms *stock* and *plug* imply a small pluton, generally something less than several miles in diameter. The term *batholith*, however, has connotations of a large pluton, exceeding 40 or 50 exposed sq mi. However, the term has been also been used to describe large areas of plutonic rocks that are pro-

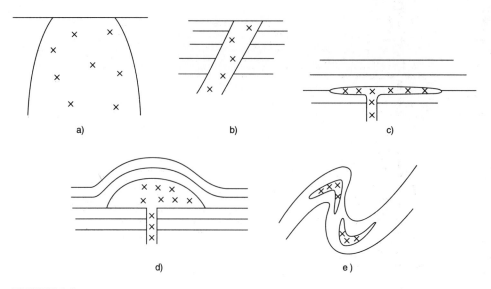

FIGURE 8.1
Geometry of plutonic bodies, with cross-sectional shapes of: a) pluton; b) dike; c) sill; d) laccolith; and, e) phacolith.

duced by several intrusive pulses; i.e., several different plutons. An example of such a batholith is the Sierra Nevada Batholith of central-southeastern California. To avoid ambiguities, the term *composite batholith* should be used for such structures.

Dikes and *sills* are tabular intrusive bodies that are generally much smaller than plutons. The distinction between them is the discordant (nonparallel) orientation of dikes within the local structure and the concordant (parallel) orientation of sills (Figure 8.2). It should be noted that in a normal horizontally bedded sequence, sills must have a *feeder dike*, the conduit for magma flow to the sill (Figure 8.1).

A sill with a domed roof and flat floor is called a *laccolith*. Like a sill it must also be connected to an underlying feeder dike. Laccoliths are the dominant structures in the Abajo Mountains of eastern Utah, and the Henry Mountains of south-central Utah, where the concept of laccoliths was first described by G. K. Gilbert.

A less common intrusive body is a *phacolith*. This lenticular concordant structure (Figure 8.1) forms in fold axes and is generally smaller than a laccolith. Its presence in the folds' axes suggests that the magma was intruded after the folds' formation. Phacoliths are prominent in the Homestake Gold Mine in the Black Hills, South Dakota.

8.2 COMPOSITE GRANITIC BATHOLITHS

Large composite batholiths such as the Sierra Nevada Batholith, the Boulder Batholith in Idaho, or the Coast Range Batholith in southern British Columbia pose an interesting question: Why are these and other large composite batholiths all of Mesozoic age? It turns out that the answer to this question can give us an insight into their internal structure. Examination of the age and geometry of the best studied composite batholith, the Sierra Nevada Batholith, can help answer this question.

The Sierra Nevada Batholith

The Sierra Nevada Batholith of central and eastern California is composed of more than 200 individual plutons, ranging in age from 210 to 70 million years. By comparison, the Boulder Batholith consists of a mere 12 or so discrete plutons. Individual plutons in the Sierra Nevada range in size from more than 100 sq mi to less than one square mile. The composition of these plutons is granitic, typically *granodiorite* or *quartz monzonite*, as well as granite.

Figure 8.3 shows an idealized east-west cross-section across the Sierra Nevada Batholith (Hamilton & Myers, 1967). The information shown in the cross-section has been derived from several geophysical methods, including gravity and seismic surveys. The cross-section demonstrates two important points: the generally assumed vertically cylindrical shape of a pluton extends downward only a few miles before it

FIGURE 8.2
Roadcut, 10 m high, exposes two diabase dikes at base, intersecting at top. Light-colored sandstone into which the dikes intruded is approximately horizontal.

becomes pinched off; and the overall geometry of a composite batholith resembles a flattened biscuit. In the case of the Sierra Nevada, this "biscuit" is elongated in a north-south direction (350 by 50 miles) and is a few miles thick. The contorted wall rocks at its margins and at pluton boundaries demonstrate that the plutons were forcefully intruded. In the process some batches of magma shown diagrammatically as isolated pods in the cross-section crystallized before they reached the general level of other intrusions. Also shown is the compositional transition from an upper *sialic* crust of granitic gneisses (rich in Si and Al) to the deeper *simatic* crust of gabbro and amphibolite (rich in Si and Mg). Note that the *Mohorivicic Discontinuity*, the lower boundary of the crust, is deeper beneath the Sierra Nevada, forming a well-defined mountain range "root."

This batholith formed in Mesozoic time above an active subduction zone dipping eastward beneath the western edge of North America (Figure 8.4). Magmas generated by partial melting from subduction moved upward through the crust and crystallized, in part, as plutons in the crust's upper levels. Some magma reached the surface and crystallized as a volcanic sequence above the granitic plutons. Based on the "biscuit" geometry of composite batholiths, their restricted age distribution can be readily explained as an erosional phenomenon. Batholiths older than about 200 million years have been eroded away. In Precambrian and Paleozoic terranes occur

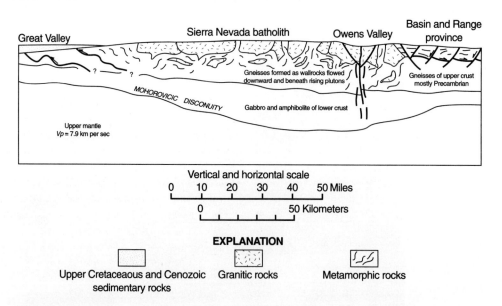

FIGURE 8.3
East-west cross-section across Sierra Nevada Batholith, eastern California, 37th parallel. Plutons of granitic magma melted in upper mantle and lower crust, rose through crust, coalesced at surface forming Sierra Nevada Batholith.
Source: Adapted from Hamilton, W., and W. B. Myers. 1967. *U.S. Geol. Surv. Prof. Pap.* 554–C, p. 5.

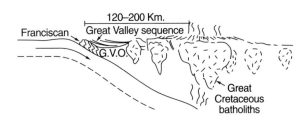

FIGURE 8.4
Subduction zone off west coast of North America in Cretaceous. Eastward-dipping subduction zone generated granitic magmas which rose through crust, forming the Sierra Nevada Batholith.
Source: Schweikert, R. A., and D. S. Cowan (1975). *Geological Society of America Bulletin*, 86, Fig. 3, p. 1334.

smaller batholiths that are merely roots of plutons formerly coalesced into a composite batholith at upper crustal levels of the crust (now eroded). The absence of composite batholiths younger than Mesozoic can be explained as resulting from erosion not yet having removed the volcanic rock cover still burying the batholiths. Further evidence is provided by the United States Geological Survey geophysical investigations in the San Juan Volcanic Field, which have revealed the outlines of an underlying batholith beneath the thick cover of volcanic rocks. Recalling the riches in gold alone produced from the Sierra Nevada in the mid-1800s, one cannot help but ponder what wonders may lie hidden within the batholiths yet to be exposed.

8.3 PEGMATITES

Structure and Mineralogy of Pegmatites

Pegmatites are plutonic rocks composed of very large crystals. Typically many crystals in pegmatites may be only an inch or two across, but, in some instances, individual crystals may have maximum dimensions exceeding 50 feet! Pegmatites typically occur as pockets, lenses, or veins in, and adjacent to, plutonic rocks. In many instances pegmatites occur within metamorphic rocks. Their compositions are variable, but the vast majority are granitic. For purposes of discussion they can be classified into one of two types: *simple* and *complex* or *exotic*. Simple pegmatites consist of quartz, alkali feldspar, and muscovite. The size of the quartz and feldspar crystals in these pegmatites is impressive, and such simple pegmatites are mined for sheet mica and feldspar. Commonly, portions of the pegmatite will exhibit an intimate intergrowth of quartz and alkali feldspar called *graphic granite* (Figure 8.5).

Exotic or complex pegmatites probably make up less than 1 percent of all pegmatites. In addition to the three minerals found in simple pegmatites are a wide array of other minerals including tourmaline, spodumene, topaz, columbite, beryl, chrysoberyl, lepidolite, and a large number of rare earth-bearing minerals. This mineral assemblage is not usually found in a single pegmatite, suggesting that exotic pegmatites are rich in rare-earth elements and lithium, beryllium, fluorine, and chlorine. A pegmatite will usually be enriched in only one or two such elements.

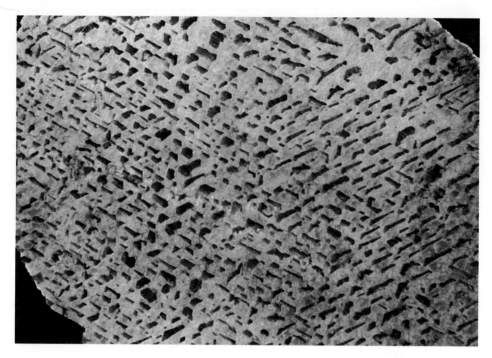

FIGURE 8.5
Graphic granite, composed of dark quartz in a host of light-colored alkali feldspar. From Llano County, Texas. Width of photograph, 6 cm.

An example of a complex pegmatite is the Harding Pegmatite in Taos County, New Mexico. It exhibits typical characteristics of large pegmatites such as a lenticular shape and internal zonation. This pegmatite was intruded in Precambrian amphibolite and quartz-muscovite schist of the Picuris Range, a branch of the Sangre de Cristo Mountains in northern New Mexico. It consists of a series of lenticular dikes (Jahns & Ewing, 1976), the largest of which is approximately 250 ft. thick and has an exposed length of more than 1,000 feet. It has been dated at about 1.3 billion years of age. The pegmatite's bulk composition is granitic. Fluid inclusion studies indicate that it probably formed at a temperature of about 600°C (Brookins et al., 1979). Eight major zones have been recognized. Toward the periphery are quartz, beryl, perthite, aplite, and quartz-spodumene zones. In the core are found the cleavelandite, rose muscovite-cleavelandite, and microcline-spodumene zones. The six most abundant minerals are quartz, albite (cleavelandite variety), microcline, muscovite, lepidolite, and spodumene. The four principal accessory minerals are beryl, garnet, microlite, and columbite. The combination of major and accessory minerals indicates that although the pegmatite's composition is granitic (quartz and feldspar rich), the dominant exotic elements are lithium (in spodumene and lepidolite) and beryllium (in beryl).

Pegmatite Formation

Pegmatite genesis has long been a source of controversy among petrologists (geologists who study rock compositions and origins). Any acceptable theory of pegmatite formation must explain the presence of very coarse-grained material in contact with medium- or fine-grained igneous or metamorphic rock—often of very different compositions from that of the pegmatite—and their commonly zoned internal structure. Proposed theories of pegmatite origins can be grouped into two major categories: igneous or metamorphic. Proponents of an igneous origin envision a hydrous silicate melt, usually granitic in composition, coexisting with an aqueous-rich fluid exsolved from the silicate melt (Jahns & Burnham, 1969). This fluid is probably rich in boron, phosphorus, chlorine and fluorine, and water. The addition of water and other volatiles lowers liquidus and solidus temperatures and the viscosity of the fluid, and thus enhances cation diffusion. Such a fluid would promote the growth of very large crystals. It could also be injected into the rock fractures surrounding an adjacent pluton, producing upon crystallization a pegmatite with a composition differing from that of the host rock.

Pegmatites occurring in metamorphic rocks are of two types: those in which the bulk composition differs greatly from the surrounding metamorphic rock, and those having a mineralogical composition similar to that of the surrounding rock. Proponents of a metamorphic origin suggest that partial melting of crustal material can yield a water-rich silicate melt of granitic composition. If this melt migrates into a portion of the crust with a different composition, it may be subsequently altered by diffusion of constituents from the surrounding rock into the crystallizing pegmatite. Pressure and pressure gradients are suggested partial causes of variation in pegmatite zonation, mineralogy, and ion diffusion from the surrounding rock.

Although petrologists on either side of the question may argue about the details of the theories, it seems reasonably certain that some pegmatites are igneous in origin and others resulted from metamorphism. The cause of the very large crystals in both pegmatite origins is an aqueous phase of much lower viscosity than a granitic melt.

8.4 STRATIFORM COMPLEXES

Gabbroic layered intrusions, or *stratiform complexes*, are large bodies of differentiated basaltic magma, which in addition to their layering, exhibit other sedimentary-like structures. They occur on all continents and range in age from Precambrian to Tertiary (Appendix I). Examples of these intrusions include the Skaergaard intrusion in Greenland and the Stillwater complex in Montana (Figure 8.6). One of the best studied and largest of these is the Bushveld Complex of South Africa (Gruenewaldt et al., 1985), Precambrian in age, and covering an area of more than 25,000 square miles, with a total volume of more than 20,000 cubic miles. In map view, the Bushveld complex consists of five lobes whose semicircular shape suggests that meteorite impacts may have been responsible for the bowl-shaped basins which it

FIGURE 8.6
Gabbroic anorthosite from central portion of Stillwater Complex, southern Montana. Light-colored areas are plagioclase, dark grains are olivine and pyroxene. Width of photograph, 6 cm.

occupies. The igneous rocks are layered on a scale varying from inches to several hundreds of feet, and dip inward at angles of about 20 degrees toward the basin center. The lower boundary of this intrusion is rimmed by a fine-grained *chill zone* recording the magma's basaltic composition prior to differentiation. As the large volume of magma cooled, minerals crystallized from it according to Bowens Reaction Series (Appendix III), and, depending on their specific gravities, either sank to the bottom of the magma chamber or floated toward the top.

The layered igneous rocks comprise the Layered Series, which has a total thickness of more than 25,000 ft and is subdivided into several zones. The lower Basal Zone is rich in olivine and pyroxene as well as the accessory mineral *chromite*. Upward in the series, the percentage of dense ferromagnesian minerals decreases while that of plagioclase increases. The lower portions of the complex are mined for chromite. The Merensky "Reef," found about 7,000 feet above the base, is a major source of platinum. In the complex's upper level, magnetite replaces chromite as the dominant iron oxide. Similar layering and zonation is found in all stratiform complexes, and layers which may be only a few inches thick are remarkably persistent and traceable for tens of miles.

8.5 ALPINE-TYPE ULTRAMAFIC BODIES

Ultramafic rocks are very rich in iron or magnesium and poor in silica and aluminum. As a result, their mineralogy is dominated by olivine and pyroxenes, and lacks feldspars and quartz. Two occurrences of ultramafic rocks have already been noted: in the basal portion of stratiform complexes, and as inclusions in basalts and kimberlites. They also form Alpine-type ultramafic bodies occurring within orogenic belts, and in concentrically zoned dunite-peridotite (Alaska-type) bodies. This latter occurrence is apparently associated with subduction zones and/or island arcs (Moores, 1973). Ultramafic rocks have a relatively simple mineralogy of pyroxenes and olivine, with accessory spinel and chromite. Some Alpine-type intrusions are predominantly serpentine. These range from building-size to more than 30 miles in diameter. Their geometry ranges from lenticular to cylindrical to irregular. Alpine-type ultramafic bodies occur in the North American Cordillera and are abundant in the Appalachians. Typically they exhibit a sheared, granular texture and are bounded by faults. These features suggest that they are fragments of suboceanic mantle that have been thrust onto the continental crust at a subduction zone. These bodies are found within both high- and low-grade metamorphic rocks and their mineralogy results from metamorphism. The conversion from serpentine to olivine occurs at approximately 500°C to 550°C. As a result, serpentine is the dominant mineral where these bodies occur in low-grade metamorphic terranes, but olivine dominates where the body is in a high-grade terrane.

8.6 ANORTHOSITES

Anorthosite massifs (mountains) are restricted to Precambrian terranes; most bodies date from about 1.5 billion years ago. Anorthosite mineralogy is dominated by calcic plagioclase which composes 50 to 90 percent of the rock. Accessory minerals include pyroxene, olivine, ilmenite, and magnetite. Anorthosite bodies are relatively common within parts of the Canadian Shield, especially Labrador and eastern Quebec. These plutonic bodies range in size from few kilometers to tens of kilometers in diameter. They are commonly associated with intrusions of syenite and diorite. The origin of anorthosite bodies is problematical, but partial melting in the mantle or deep crust during a time when the geothermal gradient was much steeper is a likely possibility (Ehlers & Blatt, 1982).

8.7 CLASSIFICATION OF PLUTONIC ROCKS

Classification and description of plutonic rocks is a fairly straightforward exercise if the rock samples are fairly fresh; i.e., not weathered or altered, and the grain size is coarse enough to permit examination of the minerals with a hand lens. The format for the description of a plutonic rock is similar to that of a volcanic rock, and includes a description of the texture—including grain size, degree of crystallinity—and textural and structural descriptions such as a porphyritic texture or flow banding

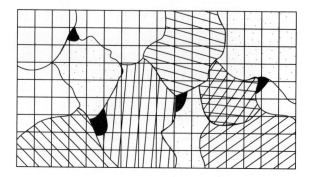

FIGURE 8.7
Determination of mode by point-counting system on symmetrical grid. The larger the number of points counted, the better the accuracy. In this example there are a total of 120 grid intersections on the rock slab: 25 (20.8%) fall on grains of quartz (clear mineral); 35 (29.2%) on alkali-feldspar (stippled grains); 35 (29.2%) on plagioclase feldspar (showing albite twinning); 22 (18.3%) on hornblende (showing 2 cleavage directions, approx. 60 and 120°); and 3 grid intersections (2.5%) on Fe-Ti oxides (black grains). The rock is hornblende granite.

(Figure 7.19). The grain size boundaries for igneous and metamorphic rocks are as shown below:

Diameter	Grain Size
≤ 1 mm	Fine
1 to 5 mm	Medium
5 mm to 3 cm	Coarse
> 3 cm	Very Coarse

A description of rock color on both fresh and weathered surfaces should be given with reference to a standard. Color is highly subjective and may be influenced by a variety of factors. For this reason, a color chart, such as the Geological Society of America Rock Color Chart, should always be used when determining rock colors. *Color index* is the volume percentage of mafic minerals in the rock. *Mafic minerals* include all those except quartz, feldspars, and feldspathoids.

Classification of Rocks Free of Feldspathoids

The first step in deriving a name is the determination of the mode. The *mode* is the volume percent composition of minerals. There are several ways to determine modal composition. The more accurate but time-consuming technique is counting mineral types at a number of points on a symmetrical grid (Figure 8.7). The accuracy improves as the number of points counted increases, but at least 100 points must be counted to reduce the error to a few percent. This technique also requires a relatively

Granites and Other Plutonic Rocks

flat surface and a method of defining a large number of regularly spaced grid points on the sample. It is the most common technique used to achieve quantitative results. A much faster, qualitative technique is visual estimation of a volume percentage for each visible mineral. With a little practice and guidance, you can determine whether alkali feldspar, for example, constitutes 10, 20, or 30 percent of the rock. A common source of error is the tendency to overestimate the percentage of dark (mafic) minerals and underestimate the amount of light-colored minerals (Figure 8.8). A good

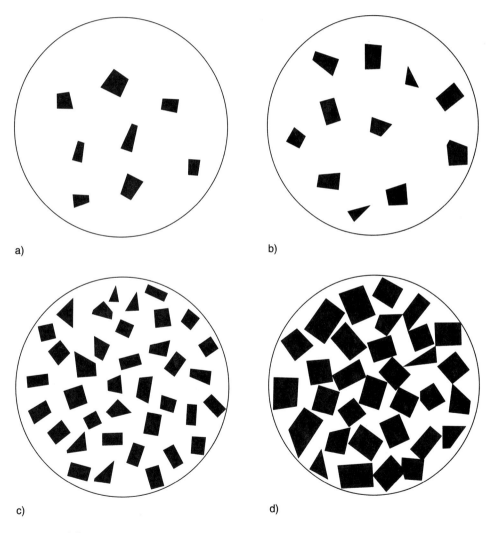

FIGURE 8.8
Diagrams to help estimate the mode. Dark grains in circles are: a) 5%; b) 10%; c) 25%; and d) 50% of total area of circle.

check on your technique is to total the estimated modal percentages and see if they add up to 100 percent. This should be done for every sample examined.

The minerals in the rock fall into three categories: essential minerals, accessory minerals, and secondary minerals. *Essential* minerals are the nonmafic minerals (except in ultramafic samples where mafic minerals constitute more than 90 percent of the mode). It is important to confidently distinguish among them and visually estimate a volume percentage for each essential mineral. This objective can be achieved by examining rock samples of known composition. *Accessory* minerals are not essential to the derivation of a root rock name. They comprise a few percent of the rock, and most commonly are biotite, pyroxenes, olivine, hornblende, and Fe-Ti oxides. Less common or abundant accessory minerals include zircon, sphene, and

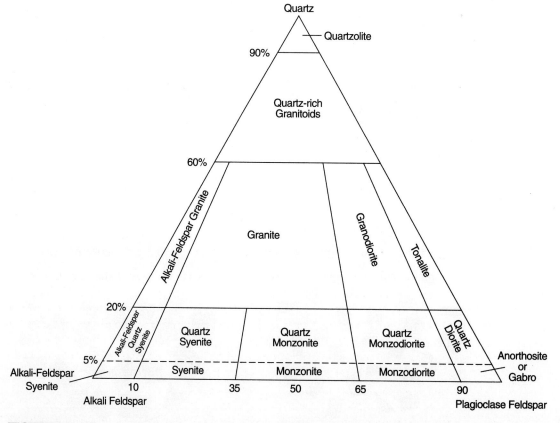

FIGURE 8.9
Classification diagram for plutonic igneous rocks according to the International Union of Geophysical Scientists.
Source: Streckeisen, A. (1976). "To each plutonic rock its proper name." *Earth Science Review.* New York: Elsevier Science Pub. v. 12, p. 8.

apatite. The *secondary* minerals result from weathering or alteration of primary minerals. If the rock is extensively weathered it may prove difficult to classify. The most common secondary minerals include hematite, clay minerals, and epidote.

The rock's root name is determined by normalizing the visually estimated percentages of the essential minerals to 100 percent, and plotting these data on the plutonic rock classification diagram (Figure 8.9). The diagram in Figure 8.9 is simply one of the latest in a long line of classification diagrams. However, it provides only a root name, not by any means a complete rock name. In order to produce a descriptive name (and accurate mental image of the rock's actual appearance), it is necessary to add modifying terms to the root name. A standard order of adjective modifiers is the following: color; grain size; characterizing accessory mineral; and lastly, root name. An example serves to illustrate the process.

A medium-grained, pale red rock has the following mode:

Percent Mineral Composition	**Modal Composition**
25% quartz	$25/90 \times 100 = 28\%$
40% alkali feldspar	$40/90 \times 100 = 44\%$
25% plagioclase feldspar	$25/90 \times 100 = 28\%$
10% biotite	—
100%	100%

Now, normalizing the essential minerals to 100 percent (by dividing the percentage of each by the summed percent of essential minerals) yields 28 percent quartz, 44 percent alkali feldspar, and 28 percent plagioclase. This composition plots in the granite field of Figure 8.9. A brief but complete descriptive rock name would be pale red, medium-grained, biotite granite.

Classification of Silica Undersaturated and Ultramafic Rocks

The following briefly describes some of the plutonic rocks that do not contain quartz as an essential constituent. *Alkalic* rocks are characterized by enrichment of alkali elements, Na and/or K, and silica deficiency. Deficiency of silica is demonstrated by the presence of feldspathoids in the mode. The most common feldspathoid is nepheline, which occurs most commonly with alkali feldspar forming nepheline syenite. Alkalic rocks have been classified in a number of ways (Streckeisen, 1976), but a descriptive classification scheme is the most sensible solution. Because most alkalic rocks contain feldspar as the most abundant essential constituent, the classification triangle in Figure 8.9 can be used to determine the root name. The presence of a feldspathoid mineral is indicated by using the mineral name as a modifier to the root name; for example, nepheline syenite or leucite gabbro.

Carbonatites are igneous rocks composed predominantly of carbonate minerals. Carbonates found in carbonatites include calcite, dolomite, and siderite. Accessory minerals may include olivine, biotite, pyroxene apatite, and nepheline. Other alka-

TABLE 8.1
Classification of Ultramafic Rocks

Mineralogy	General Name	Specific Name
>90% olivine (Ol)	Dunite	Dunite
>90 orthopyroxene (Opx)	Pyroxenite	Orthopyroxenite
>90% clinopyroxene (Cpx)	Pyroxenite	Clinopyroxenite
Mostly Opx and Cpx	Pyroxenite	Websterite
Mostly Opx and Ol	Peridotite	Harzburgite
Mostly Cpx and Ol	Peridotite	Wehrlite
Mostly Ol, but with 10-30% Cpx and Opx	Peridotite	Lherzolite

lic rocks such as nepheline syenite are commonly found associated with carbonatites.

Ultramafic rocks are composed predominantly (more than 90 percent) of mafic minerals. Except for kimberlite, ultramafic rocks are classified by the relative proportions of the following minerals: olivine, clinopyroxene, and orthopyroxene (Table 8.1). *Kimberlite* is an ultramafic rock characterized by phlogopite, but which may contain a large number of accessory minerals including garnet, olivine, pyroxenes, and serpentine. They occur in stable, nonorogenic cratonic areas. The presence of diamonds in some kimberlites, and the mineralogy and chemistry of the ultramafic xenoliths within kimberlite suggest an origin at depths of more than 150 km (Meyer, 1977). The kimberlite was probably emplaced as a gas-rich magma that rapidly bored its way to the surface in an explosive manner.

Lamprophyres are a group of rocks occurring in dikes and sills. They are characterized by large, well-formed phenocrysts of mafic minerals, typically biotite or hornblende. Feldspar minerals may be present in the groundmass (fine-grained matrix), but not as phenocrysts. Their compositions can range between gabbro and a strongly silica-undersaturated rock. Many lamprophyres contain abundant feldspathoid minerals.

9
Sedimentary Rocks I:
The Detrital Sedimentary Rocks

A sedimentary deposit results from accumulation of solid materials at or near Earth's surface at conditions of low temperature and pressure. The solid materials that accumulate can result from weathering and deposition producing a *detrital* sedimentary rock consisting of fragments of pre-existing rock cemented together. Detrital sedimentary rocks are produced secondarily by the weathering and breakdown of pre-existing rock, unlike igneous rocks, which can be derived directly from the mantle and deposited where no crustal material existed before.

9.1 WEATHERING

The first step in sediment formation, the raw material necessary for sedimentary rock formation, is weathering. Weathering affects all rock types but is especially effective on igneous and metamorphic rocks, because they formed at much higher temperatures and are therefore more susceptible to changes in the surface environment. *Weathering* is defined as the processes that alter the texture, mineralogy, and chemistry of rocks at or near Earth's surface to stable products in this low temperature and pressure environment. The minerals formed by weathering depend on a number of factors, the most important of which are mineral and chemical composition of the parent rock, climate, organic activity, and time. The original chemical composition of the unaltered rock limits the weathering products' composition. For example, the plagioclase feldspar in basalt, which contains calcium, sodium, aluminum, and silica, is likely to weather to a mixture of calcite and albite or clay minerals. Similarly, alkali feldspar ($KAlSi_3O_8$) will likely weather to kaolinite, but is unlikely to yield any calcite.

FIGURE 9.1
Bauxite sample showing typical pisolitic texture. Coin is 1.8 cm diameter.

Climate is one of the most important weathering factors because it controls water availability. Water's availability, in turn, influences the amount of biological activity. In a humid climate, where rainfall is abundant and frequent, both physical and chemical weathering are enhanced, and weathering is more rapid and pervasive. Abundant rainfall also ensures intense leaching and the presence of water encourages plant and bacterial activity. Decay of plant material produces carbon dioxide, which combines with water to produce carbonic acid (H_2CO_3), a much more effective weathering agent than pure water.

In tropical climates, the heavy rainfall and dense vegetation produce intense, deep leaching of soils. This intense leaching, combined with high year-round temperatures of tropical regions, produces, with time, a very stable assemblage of weathering residues, or soils, called *laterites*. Aluminum-rich laterites are called *bauxite* (Figure 9.1), and are common in tropical regions underlain by carbonate rocks. Mafic and ultramafic rocks produce laterites enriched in iron, nickel, or chromium.

9.2 SEDIMENTARY ENVIRONMENTS

The relative abundance of igneous and sedimentary rocks in and on the Earth's crust tells a revealing story. Although sedimentary rocks make up only 5 percent of Earth's crustal volume, they cover 75 percent of its surface. These statistics suggest that sedimentary rocks are the product of surface processes and are concentrated at or near the surface. In the near surface environment, sediments accumulate both above and below sea level, on continents and in ocean basins. Deltas are the immediate identifiable accumulations of terrigeneous sediment (eroded from continents), flushed into the ocean basins by large rivers. Some of the deltaic sediments are subsequently redistributed along the shoreline by waves and currents. In the Gulf of Mexico, for example, deltaic sediments of Cenozoic age exceed more than 50,000 feet in thickness. Lakes, streams, and the continental shelves and abyssal plains of the marine environment are other places where sediment accumulates today. Thick accumulations of sediment are also found in deep oceanic trenches, the location of active subduction zones.

9.3 STORIES THE ROCKS CAN TELL

Earth's history is written most clearly in the layers of sedimentary rocks. They record not only the transgressions (sea level increase) and regressions (sea level decrease) of ancient seas on the continents, but also contain traces of ancient life, *fossils*, revealing what creatures lived in the geologic past. Fossils record the increasing diversity

FIGURE 9.2
View downstream of Marble Canyon cut by Colorado River between Lees Ferry and Grand Canyon, Arizona. All are bedded Paleozoic sedimentary rocks. Steep cliffs near the bottom of the photograph are cut into the Mississippian Redwall Limestone.

FIGURE 9.3
Laminated gypsum from Permian Castile Formation, Delaware Basin, Texas, and New Mexico, with a coin less than 2 cm in diameter for scale.

and macroevolution of life from the Precambrian to present. One of the paradoxes of this record of Earth history is that in order to write more history, some traces of more ancient history must be erased, because deposition cannot occur without erosion. The record of today's geologic events is being written in new sediment derived by erosion of older rocks. It is as if we had a book with only a certain number of pages. The pages are all full so that in order to write in it, something that has been previously written must be erased. The great eraser is erosion. The consequence of this fact is an incomplete geologic record.

Sedimentary Structures

The key to deciphering the geologic history "written" in the sedimentary rock column is a proper interpretation of preserved fossil assemblages and sedimentary structures. The most obvious field feature of sedimentary rocks is bedding (Figure 9.2). These bedding planes are formed by short interruptions in sedimentation or by changes in sediment composition. A sequence of beds can be described as thin (less than one ft), medium (one to three ft) or thick (more than three ft): bedding is usually not visible in a laboratory hand specimen. If the bedding is very thin (<1 cm), the unit is described as *laminated* (Figure 9.3). A closer look at sedimentary rock beds usually reveals a number of clues to the sedimentary environment of deposition, in the form of primary sedimentary structures including sole marks, cross-bedding, ripple marks, and mudcracks. All these features can be used to determine the top and bottom of beds, and thus, whether or not the sequence is in its original horizontal position.

Sole marks are filled-in depressions in the underlying bed of sediment, usually sandstone or siltstone. Some of these depressions are scour marks made by debris bouncing along the bottom while transported by a wave or current. These marks provide a clue to the current direction during deposition. *Flute casts* (Figure 9.4) are asymmetrical structures on the bottom of sandstone beds produced by scouring and erosion of fresh mud by a *turbidity current* (dense, bottom-hugging flow of sediment-laden water from submarine slumping). The structure's asymmetry enables determination of the current direction—the deepest part of the flute indicating the upcurrent direction.

Cross-bedding is a group of similar structures produced by downslope movement of sand or silt resulting in thin, often curved beds formed at an angle to the larger, horizontal bedding planes (Figure 9.5). Cross-bedding is usually best seen in cross-sections of beds rather than on bedding planes. The crossbed are usually

FIGURE 9.4
Sole marks on bottom of sandstone bed; early Paleozoic age, west coast of Newfoundland.

FIGURE 9.5
Cross-bedding in Jurassic sandstone, northern Arizona.

curved, being steeper toward the top and approaching horizontality near the bottom. Crossbeds form horizontal or near-horizontal erosionally bounded packages called *sets*. The crossbeds range in size from ripple marks to sand dunes and are produced by currents or waves of water or wind. Cross-bedding can be used to interpret the physical conditions of a detrital sedimentary rock's formation, including water velocity and relative depth, and direction of sand transport at the time of deposition. Some large-scale cross-bedding (Figure 9.5) results from wind currents producing sand dunes in arid environments. Others are products of high energy currents in shallow marine environments.

Ripple marks are the surface expression of small-scale cross-bedding. They can be produced by currents and waves and their morphology differs with each mode of origin (Figure 9.6). Current-produced ripple marks are asymmetrical with a steeply dipping limb on the current's lee side, while wave-generated ripple marks are more symmetrical, reflecting the symmetrical wave form.

Mudcracks are another feature found on bedding planes. They are a good indicator of the "up direction" in a sedimentary sequence and also provide clues to the environment of deposition. Because mudcracks are produced by drying of wet sediment, their presence suggests an environment subjected to repeated dry and wet cycles, most likely at, or close to, sea level where there is periodic submergence by tides, as occurs in a tidal flat.

Allogenic and Authigenic Constituents

More intensive study of detrital sedimentary rocks reveals a dual story line, based on the two components of which they are comprised. The *allogenic* component, that which has been transported to the place where it is found, tells us something about the material's source terrane. The *authigenic* component, which forms at the deposition site, includes the cementing material and provides a clue to the depositional environment. The most reliable way of distinguishing between the allogenic or detrital component and the authigenic material is the rounding of the allogenic component by erosion (transportation). The detrital component usually consists of two mineral groups, those forming a major portion of the rock, such as quartz and feldspar, and those present in trace amounts. This latter group includes "heavy minerals," so-called due to their above-average specific gravity. They are important in determining the sediment's provenance or source rock(s). Common heavy minerals include zircon, garnet, rutile, and tourmaline.

By utilizing the information in allogenic and authigenic components of detrital sedimentary rocks it is possible to not only determine the environment and paleogeography (ancient geography) of the immediate area of deposition, but also the paleogeography of more distant source areas, for example, the location of, and distance to, an ancient mountain range.

FIGURE 9.6
Ripple marks in siltstone of Belt Group, Late Precambrian, western Montana. Note presence of mudcracks.

Diagenesis

Diagenesis includes all physical and chemical changes occurring in sediment after deposition and prior to metamorphism. Several distinct processes are encompassed in this definition, including compaction, dewatering, cementation, dissolution, crystallization of minerals such as zeolite and dolomite, and lithification. Many of these processes are still not completely understood but diagenetic processes affect rock porosity and permeability, and its color, hardness, and chemical composition. The result of this mix of processes is the conversion of sediment into sedimentary rock.

9.4 DESCRIPTION AND CLASSIFICATION OF CLASTIC SEDIMENTARY ROCKS

As with all rocks, detrital sedimentary rock classification is a function of texture and composition. Also, a number of physical properties describing the detrital component are integral to the description, including maturity, sorting, and roundness.

Maturity is the degree to which a sedimentary process approaches completion. Two types of maturity are recognized: *textural maturity* and *compositional maturity*. A texturally mature sediment is one consisting of well-sorted and well-rounded grains, in which the range of particle diameter is limited, and in which particle corners have all been knocked off. A compositionally mature sediment is one consisting of minerals stable at 25°C and 1 atm. An example of a compositionally and texturally mature sediment is a sandstone composed of well-sorted and well-rounded grains of quartz, with minor zircon. An example of an immature sediment is one composed of angular (texturally immature) fragments of augite and calcic plagioclase (compositionally immature). Neither calcic plagioclase or augite are stable at 25°C. They occur in detrital sediment only if it is relatively young or has been buried before weathering altered the minerals. As a general rule, the higher a mineral's temperature of formation, the less stable it will be at 25°C. A long list of minerals has been documented as forming in sedimentary environments, but the most common and abundant minerals are quartz, albite, alkali feldspar, and clay minerals.

Sorting is a measure of grain size variability (Figure 9.7). It always appears more variable on a cut surface or a thin section because the planar surface truncates

FIGURE 9.7
Diagram used in estimating degree of sorting: a) well-sorted, and b) poorly sorted.

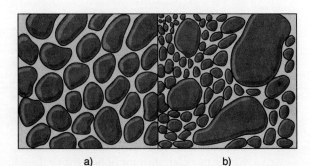

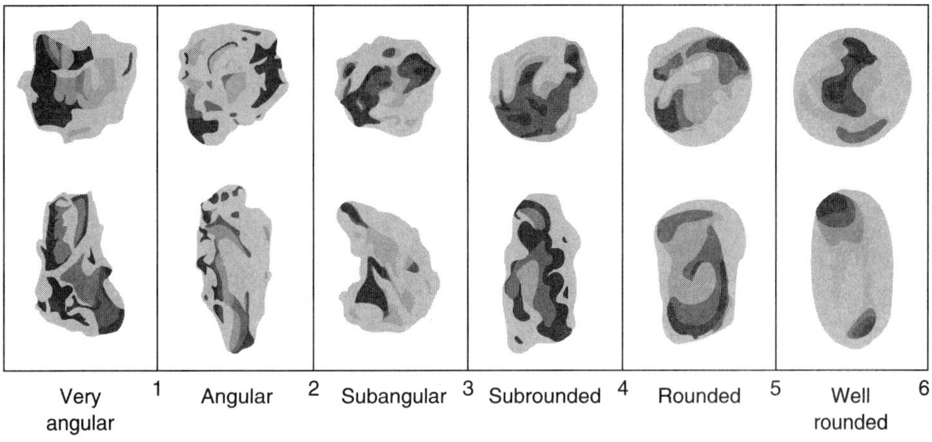

FIGURE 9.8
Qualitative estimates of angularity and roundness for sand-sized particles.
Source: Powers, M. C. (1953). *J. Sed. Pet.* v. 23, Fig. 1.

grains along various cross-sections, thus exaggerating the size variation. The best place to estimate the degree of sorting is along a natural fresh break. Most sandstones exhibit moderate to good sorting because the water or wind currents that deposited them have winnowed the smaller grains. Most glacial sediments are poorly sorted because transportation by ice does not discriminate on the basis of grain size.

Roundness refers to the degree of curvature of grain surfaces. This parameter is independent of *shape*, and used to describe grains of sand size or larger. Although techniques for metric measurement of roundness have been devised, a qualitative estimation, such as well-rounded or angular, suffices for most descriptions (Figure 9.8).

The Classification of Detrital Sedimentary Rocks

The starting point for detrital rock classification is determination of grain size. A variety of grain-size scales have been proposed but the most widely accepted is The Wentworth (1922) Size Scale (Figure 9.9). In common terms, texture can be defined as coarse, medium, fine, or very fine: the equivalents of gravel, sand, silt, and clay, respectively. In some classification systems the Latin-derived terms *rudite*, *arenite*, and *lutite* are used for gravel, sand, and clay, respectively.

Determination of root names for sedimentary rocks is slightly different for the three size ranges. Anything coarser than a sand is a *conglomerate* (rounded fragments, Figure 9.10) or a *breccia* (angular fragments). Consolidated sediments composed of sand-sized grains (sandstone) are classified on the basis of mineral and textural components. Anything finer grained than a sandstone is known as a *mudrock*, consisting of either clay or silt, or a mixture of the two. Mudrocks are classified by the

FIGURE 9.9
Wentworth size scale for clastic sedimentary rocks.
Source: Wentworth., C. K. (1992). *J. Geology.* v. 30, Table 1, p. 381.

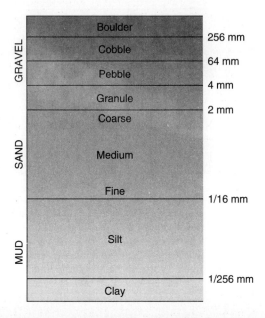

FIGURE 9.10
Conglomerate composed of rounded cobbles and boulders cemented in coarse sand matrix. From Arkansas River valley, central Colorado. Most cobbles and boulders are clasts of igneous or metamorphic rocks. Large boulder in center is 25 cm in longest dimension.

FIGURE 9.11
Sample of black shale, showing typical fissile character. Sample is 10 cm wide.

relative abundance of the clay and silt fractions and the presence or absence of fissility (Table 9.1 and Figure 9.11).

Sandstones include the largest assortment of names and classification triangles (Klein, 1963). An example of such a triangle is shown in Figure 9.12. The essential constituents of sandstone classification are quartz, feldspar (usually alkali feldspar), and rock fragments. Other detrital components are rare in common types of sandstone. *Wackes* are a group of sandstones characterized by the presence of large amounts (15 to 75 volume percent) of fine-grained matrix. Wackes are abundant in basins formed near orogenic belts, and they are generally texturally and compositionally immature.

The first step in detrital rock classification is determining whether the rock is a conglomerate or breccia, sandstone, or mudrock. The second step is refining the root name by determining the mineralogy or lithology of the allogenic and authigenic components. Three examples illustrate the classification sequence:

> *Example 1*—A conglomerate composed of rounded chert pebbles and cemented by hematite. Its color, determined by comparison with the rock color chart, is light brown. A complete brief name would thus be a light brown, hematite-cemented, chert pebble conglomerate.

TABLE 9.1
Classification of Mudrocks

Ideal size definition	Field criteria	Fissile mudrock	Nonfissile mudrock
> 2/3 silt	Abundant silt visible with hand lens	Silt–shale	Siltstone
1/3 < silt < 2/3	Feels gritty when chewed	Mud–shale	Mudstone
> 2/3 clay	Feels smooth when chewed	Clay–shale	Claystone

Source: Blatt, et al., 1980, *Origin of Sedimentary Rocks*, 2nd ed., Table 11.1.

Example 2—A reddish-brown siltstone having laminations. This rock would be described as a reddish brown laminated siltstone.

Example 3—A very light gray sandstone composed of well-rounded, medium-grained quartz (95 percent), minor feldspar and heavy minerals, and calcite cement: the rock is called a quartz arenite (Figure 9.12). A brief but complete name would be light gray, calcite cemented, medium-grained quartz arenite.

These examples give a glimpse into sedimentary rock classification methodology. Note that the classification implies no interpretation of their origins.

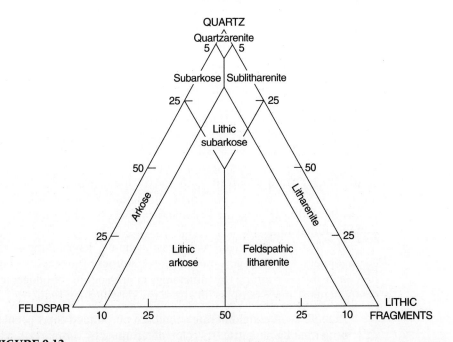

FIGURE 9.12
Sandstone classification triangle
Source: McBride, E. F. (1963). A classification of common sandstones. *J. Sed. Pet.* v. 33, Fig. 1, p. 664.

9.5 GENETIC INTERPRETATIONS OF SOME COMMON SANDSTONES

Sandstones are composed of various source materials. Some of the more common source terranes include volcanic deposits, producing a volcaniclastic sand or *pyroclastic* sand; chemical sedimentary rocks producing some limestones or *chert lutites;* a metamorphic source yielding *phyllarenites;* and *terrigeneous sands* composed of quartz, feldspar, and rock fragments. The most abundant sandstones in the geologic column are of terrigeneous origins, of which arkose and quartz arenites are the most abundant members.

Quartz Arenite

Quartz arenite is characterized by more than 95 percent quartz. It is more texturally and compositionally mature than any other sandstone. The quartz grains are typically well-rounded and well-sorted. The remaining 5 percent of the rock may be composed of resistant heavy minerals such as zircon, rutile, or tourmaline. The presence of stable minerals in these sandstones indicates a period of tectonic stability during deposition. Weathering and erosion over millions of years produces a sediment composed of quartz grains, the most resistant common mineral. The usual cement is silica but may be calcite. Small- or large-scale cross-bedding is prominent in some quartz arenites. The presence of large-scale cross-bedding in some is evidence of an eolian (wind-blown) origin. Fossils are rare but fragments and trace fossils may be present.

Some of the better known formations consisting of quartz arenite are the Ordovician St. Peter Sandstone, which covers a large portion of the Mid-Continent region, and the Permian Coconino Sandstone of the Grand Canyon region.

Arkose

Arkose is characterized by a substantial amount (more than 25 percent) of feldspar. This feldspar is usually microcline and is typically partially weathered to kaolinite. The most abundant mineral is quartz, comprising less than 75 percent of the rock. Flakes of muscovite are a common accessory mineral. Calcite is the typical cement. Unlike texturally mature quartz arenites, arkoses are immature sandstones. The degree of immaturity is evinced by the kaolinized feldspar, angular and poorly rounded grains, and poor sorting. These characteristics suggest that arkoses are deposited fairly close to their source terranes and rapidly buried. The two common occurrences of arkose further support this interpretation.

Arkose occurs as thin basal layers directly overlying granitic terranes. In many instances, the sandstone is conglomeratic or interbedded with granitic conglomerate clasts. The volume of these arkose deposits is generally small, grading upward into more widespread quartz-rich arenites.

Arkose also occurs as wedge-shaped deposits adjacent to uplifts of granitic rock. These deposits, also conglomeratic adjacent to the uplifts, grade into finer

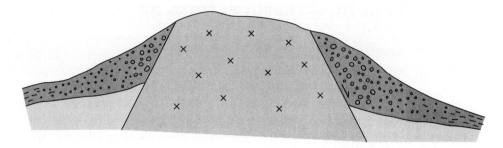

FIGURE 9.13
Wedge-shaped deposits of arkose and arkosic conglomerate adjacent to granitic block uplift.

grained sandstone and siltstone with increasing distance from the source area (Figure 9.13). An example is the Fountain Formation of Pennsylvanian age formed adjacent to the Ancestral Rocky Mountains in central Colorado.

The occurrences of arkose support the idea evident upon first glance—that there is a close relationship between granitic rocks and arkoses. In fact, in some specimens, arkose may be mistaken for granite. Yet, although granites are widespread and abundant on continents, arkose is not an abundant sedimentary rock. The reason is that arkose formation requires, in addition to a granitic source terrane, rapid burial and a climate favorable to feldspar preservation. The presence of a wedge-shaped arkose deposit, unlike the tectonic quiescence represented by a quartz arenite deposit, suggests a tectonically active period in a granitic source area. The Triassic Newark Group of fault-bounded basins in the eastern United States is a good example of an arkose.

Lithic Arenites

As the name implies, *lithic arenites* are rich in lithic (rock) fragments. Sometimes they are called "salt-and-pepper" sandstones, because of their speckled appearance. Lithic arenites can be derived from various source terranes. Because they are composed of lithic grains, each of which is composed of more than one mineral, the percentage of lithic grains in a sandstone increases with increasing sand size, or a decreasing source rock grain size. It is unlikely that a source terrane of granitic igneous rock or coarse-grained metamorphic gneisses and schists will yield many lithic grains. The most likely source terranes of lithic grains are fine-grained, generally low-grade, metamorphic rocks (slates and phyllites), volcanic rocks, and fine-grained sedimentary rocks.

Lithic arenites are among the most abundant of sandstones. They are especially abundant in the Tertiary Gulf Coast sequence and Paleozoic Appalachian sequence. Most lithic arenites are immature.

Other Sandstones

A number of terms lying outside the formal sandstone classification scheme are used informally to describe less common sandstone types. The term *placer sand* has already been mentioned in the discussion of minerals. Placer sand is an economic term describing economic accumulations of one or more sand-sized minerals. Many are sediment deposits rather than lithified sandstones. Placer sand may be mined for gold, chromite, ilmenite, zircon, rutile, or monazite.

Greensand is a term used for glauconite-bearing marine sandstone. True to the term, these sands may range in color from a light- to dark-green, depending on the percentage of glauconite. Glauconite is a green authigenic, iron-bearing, phyllosilicate mineral occurring as pellets. It may be a weathering product of biotite, which its chemical composition closely resembles. Glauconite is especially abundant in Cambrian sandstones.

9.6 MUDROCKS

Mudrocks are composed of all sediment finer than sand. The term *clay* is commonly used as a particle size term (<1/256 mm in diameter) *and* as the name for a mineral group (the former use being informal in most sedimentary rock classifications). More than 50 percent of the sedimentary rock column is composed of mud. A simple classification scheme for mudrocks is presented in Table 9.1. Fissile mudrocks are called *shale*. *Fissility* is a characteristic causing the rock to separate along thin planes parallel to bedding. The degree of fissility depends on several factors, including composition and, most notably, the amount of the mud's disturbance by burrowing organisms. Mudrocks accumulate in deltaic environments, on the continental shelf, and in deep oceanic environments.

Nonfissile mudrocks are classified on the basis of the most abundant particle size in the rock, e.g., siltstone, mudstone, or claystone. Color is an essential part of mudrock description, e.g., black shale or red siltstone.

Clay Mineralogy of Mudrocks

The three most abundant clay minerals in sedimentary rocks are illite, montmorillonite, and kaolinite. The relative proportions of these three clay species in rocks varies with geologic age (Figure 9.14). *Montmorillonite* is most abundant in Tertiary mudrocks, comprising as much as 60 percent of the clay component. *Illite* is most abundant in older rocks, making up about 80 percent of the clay fraction in early Paleozoic rocks (Ehlers & Blatt, 1982). One possible explanation for this distribution is that diagenetic reactions transform montmorillonite to illite.

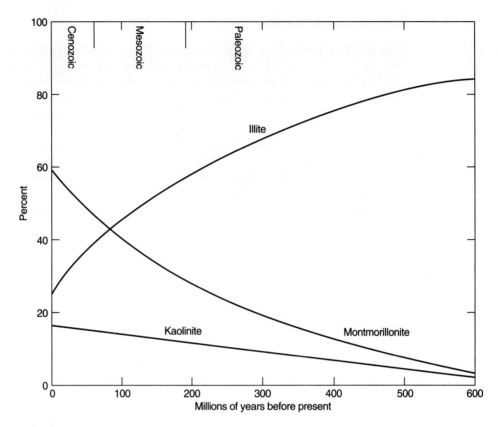

FIGURE 9.14
Distribution of different clay minerals in sedimentary mudrocks of different ages.
Source: Ehlers, E. G., and H. Blatt (1982). *Petrology*. San Francisco: W. H. Freeman. Fig. 12–7, p. 294.

10
Sedimentary Rocks II:
The Chemical and Organic Sedimentary Rocks

Chemical sedimentary rocks, in contrast to detrital sedimentary rocks, form by precipitation from a solution. The solution can be of freshwater origin (lake, spring, or river water), but the most abundant solution at Earth's surface is sea water. Therefore, most chemical sedimentary rocks are the products of precipitation from brines derived from seawater. An examination of the chemistry of seawater will enable a prediction of the kinds of minerals likely to precipitate from it. Most sedimentary rock volume formed of organic remains (e.g. limestones composed of invertebrate shells and skeletons and algae aragonite needles, and cherts formed of protistan test), except coal, are also of marine origin.

10.1 CHEMICAL COMPOSITION OF SEAWATER

Although almost every element in the periodic table is present in seawater, most are present in only trace amounts. The bulk of the material dissolved in seawater consists of seven ions: Na^+, Mg^{2+}, Ca^{2+}, K^+, Cl^-, SO_4^{2-}, and CO_3^{2-} (Table 10.1). Actually, the dissolved carbon component exists in a number of compounds including carbonate and bicarbonate (HCO_3^-). Most of the ions present in seawater are contributed by rivers and processes that selectively remove ions from solution, such as precipitation and adsorption.

TABLE 10.1
Classification of Chemical and Organic Sedimentary Rocks

Particle Size	Mineralogy	Distinguishing Characteristics		Rock Name
Fine to Coarse-Grained	Calcite	fizzes with dilute HCl	abundant fossils	Fossiliferous Limestone
			fine-grained	Fine-grained Limestone
			coarse-grained	Crystalline Limestone
			powdery-shells of microscopic animals	Chalk
			mixture of calcite and clay, usually fossiliferous	Marl
			mostly macroscopic shells and fragments	Coquina
	Dolomite	powdered rock fizzes with dilute HCl		Dolostone
	Halite	tastes salty (NaCl)		Rock Salt
	Sylvite	tastes bitter (KCl)		Sylvite
	Gypsum	hardness of 2, usually granular		Gypsum
	Chalcedony or Chert	hardness of 7, will scratch glass, conchoidal fracture		Chert
	Carbonaceous Material	brown, porous, with plant fossils		Lignite Coal
		black and blocky, brittle		Bituminous Coal

If seawater evaporates completely, the bulk of the material precipitated will consist of a variety of sulfates, carbonates, and chlorides of Na, Mg, Ca, and K. However, only rarely is seawater completely evaporated. More commonly seawater is partially evaporated, so it is important to determine exactly which minerals will form upon evaporation and their order of precipitation as evaporation proceeds.

Order of Precipitation

Precipitation of a mineral from any solution occurs when the concentration of the ions forming the compound is greater than the solubility of the compound in the solution. When the concentration equals the solubility, the solution is said to be *saturated*, for that compound. Many factors can affect the solubility of, for example, halite (NaCl) in a solution. Some of these include temperature, presssure, pH, and the concentration of other ions in the solution, such as dissolved CO_2. Seawater is a relatively homogeneous solution in the open ocean because winds and currents keep it well mixed. However, there are large variations in temperature, pressure, and dissolved CO_2 between the oceans' surface and the bottom. The solubility of NaCl in seawater is much higher than the amount of dissolved sodium and chloride, thus, seawater is undersaturated with respect to halite. Calcite is the only mineral saturated, or even slightly supersaturated, in seawater.

If a sample of seawater evaporates, the first compound to precipitate would be calcite (in trace amounts). This precipitation would be begin almost immediately because seawater contains as much calcium and carbonate as it can hold in solution, i.e., it is saturated. Further evaporation would promote the precipitation of gypsum ($CaSO_4 \cdot 2H_2O$) or anhydrite ($CaSO_4$). At the point where the seawater solution is one-tenth of its original volume, halite begins to precipitate. Just before complete evaporation, Mg- and K-chlorides precipitate.

After the solution has evaporated, the most abundant mineral left is halite, constituting more than 68 percent of the total. The second most abundant mineral groups are potassium and magnesium chlorides [sylvite (KCl), epsomite ($MgSO_4 \cdot 7H_2O$), carnallite ($KMgCl_3 \cdot 6H_2O$)], making up almost 28 percent of the total. Gypsum makes up 3 percent and calcite, the first to precipitate, makes up less than 1 percent. It should be noted that the solubility of the mineral species is more important than its ionic concentration in determining the order of precipitation from seawater. All of the minerals formed in this evaporation experiment, with the exception of calcite, comprise a group of common chemical sedimentary rocks called *evaporites* (discussed later in this chapter).

10.2 LIMESTONE

Limestone, composed principally of calcite, forms primarily in the marine environment. In this environment, calcite and aragonite accumulate in warm, shallow water areas such as off the coast of Australia, Jamaica, Bermuda, Florida, and the

Bahama Bank. In the South Pacific it forms numerous coral reefs or coral atolls. Unlike many other compounds, calcite's solubility decreases as the solution temperature increases. This decreased solubility promotes precipitation. As a result of this temperature solubility relationship, limestone accumulations do not occur in cold waters. At high latitudes and depths greater than about 13,000 feet (the "carbonate compensation depth") calcite dissolves because cold seawater's capacity to hold calcium carbonate in solution is increased (Press & Siever, 1985). However, although calcium carbonate precipitation occurs in some instances in warm shallow marine environments, the bulk of carbonate rocks forming today and those in the geologic record formed by biochemical precipitation.

Outside the marine environment, calcite can accumulate in warm water lakes, springs, and caves where it is the main constituent of cave decoration or formations. Some of the $CaCO_3$ precipitated in these environments forms as the mineral aragonite. Calcite accumulations are a problem in water heaters and hot water pipes in regions where the water contains significant amounts of dissolved calcium carbonate. The common thread in all of these occurrences is the tendency of calcium carbonate to precipitate in warm environments in which water is saturated with respect to calcite or aragonite.

FIGURE 10.1
Portion of Cretaceous reef limestone, central Texas, showing shelled organisms constituting the limestone. Width of photograph, 8 cm.

Sedimentary Rocks II

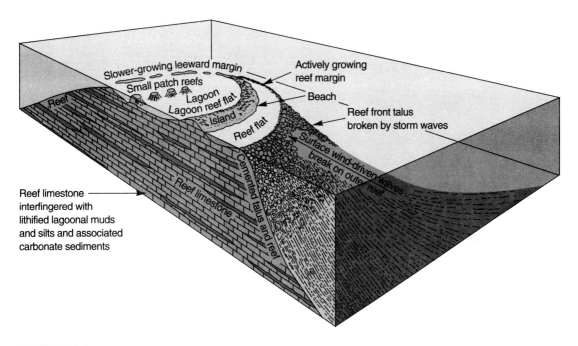

FIGURE 10.2
Map and cross-section of coral reef. Reef tends to grow more rapidly on windward side where waves provide nutrients.
Source: Press, F., and R. Siever (1985). *Earth*. 4th ed. San Francisco: W. H. Freeman. Fig. 12–7, p. 316.

Biochemical Precipitation

A variety of organisms secrete calcium carbonate to form their shells and skeletons. These include gastropods (snails), oysters, and other varieties of bivalves (clams) (Figure 10.1), brachiopods and corals (especially colonial reef-formers). It is estimated that nearly all of the present carbonate sediment volume is produced by biochemical precipitation. Two current types of environments where biochemical carbonate precipitation is common and abundant will be described: coral atolls of the South Pacific and carbonate platforms or banks.

Coral Atolls   Charles Darwin, during his epic voyage on the *Beagle* (1831 to 1836), studied coral atolls in the South Pacific and correctly interpreted their origin. The essential features of these atolls include: a) an outer, wave front absorbing ocean wave impact; b) a flat reef platform; and c) a shallow lagoon behind or within the wave front (Figure 10.2). The essential reef front is the living, growing portion lying within about 60 feet of the surface. Corals are among many invertebrate animals common in warm, shallow, marine environments. Most corals form colonies and all produce hard external calcium carbonate coverings. Certain algae species and bryozoans also thrive in coral reefs. The living portion of the reef, as Darwin discovered,

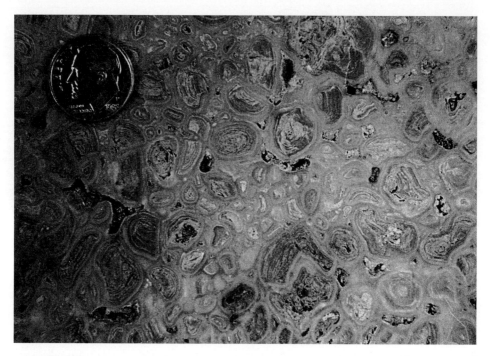

FIGURE 10.3
Pisolitic carbonate rock of Permian age, Guadalupe Mountains, southeastern New Mexico.

is built on older, nonliving reef material. However, at one time it, too, must have occupied that critical depth zone receiving enough sunlight to grow. How, then, did these atolls originate in waters thousands of feet deep? Darwin hypothesized that these accumulations of coral built upon the tops of ancient, now extinct volcanoes. As these volcanoes, and the lithospheric plate of which they are a part, cooled and subsided, the volcano top sank below sea level. However, its subsidence was slow enough that coral growth kept pace. In many cases there is little evidence of a volcano at the surface.

Carbonate Platforms Carbonate platforms or banks are areas of warm shallow water that receive a very small amount of terrigenous sediment. In these warm, clear, shallow waters, biochemical and inorganic carbonate precipitates subsuming the minor volume of detrital sediments. One of the more striking examples of inorganic precipitation are small spherical structures called *oölids*. Less than 2 mm in diameter, and with a concentric structure, they are interpreted to form around an organic fragment or silt particle acting as a nucleus around which layers of aragonite precipitated. When lithified, these particles are known as oölids. In order to ensure even coating, the particles must have been continually agitated during formation by currents or waves. Slightly larger (1–10 mm diameter), but similar in structure, are the spherical objects called pisolites. *Pisolites* are concentrically laminated structures (Figure 10.3) that resemble oölites, but are now considered the result of diagenesis (Blatt et al., 1980). The bulk of carbonate shelf sediment lies in the mud to sand-size range (Folk, 1974). The sand-sized fraction includes fragments of invertebrate skeletal remains, calcareous algae, oölites, pellets, and larger structures such as oncolites and intraclasts. *Pellets* are round to oval silt- to sand-sized invertebrate fecal particles with no internal structure, and may be replaced by microcrystalline carbonate (Blatt et al., 1980). *Oncolites* are small (cm sized) spherical accumulations of calcium carbonate precipitated by the effects of algae producing small structures with contorted layering. *Intraclasts* are fragmented plates of lithified carbonate mud eroded from tidal flats and shallow water deposits while still plastic, transported a short distance, and redeposited. Intraclasts typically are several centimeters in maximum dimension (Figure 10.4).

Carbonate mud is derived by the mechanical disintegration of larger skeletal and shell carbonate particles, and to a lesser extent from the calcium carbonate precipitation from seawater, mostly in the form of aragonite (Blatt et al., 1980). Some examples of carbonate platforms are the Florida and Yucatan peninsulas and the Bahama Islands (Figure 10.5).

FIGURE 10.4
Intraclastic limestone. Intraclasts ripped up from nearby source, transported a short distance, and redeposited.

FIGURE 10.5
Calcium carbonate debris cemented by calcium carbonate, Yucatan Peninsula, Mexico. Width of photograph, 6 cm.

10.3 DOLOSTONE

Dolostone, also less formally called *dolomite*, is composed mostly of the mineral dolomite, $CaMg(CO_3)_2$. Unlike limestone, dolomite is rarely observed forming in present marine or freshwater environments. However, older sedimentary rocks are much richer in dolomite than Mesozoic and Cenozoic rocks and recent sediments. More than one-fourth of Paleozoic carbonate rocks and more than one-half of the Precambrian carbonate rocks are dolostone.

Many carbonate formations contain both dolostone and limestone. The origin of dolostone has been a raging controversy for many decades but the present day concensus is that it is formed by some type of replacement of limestone by dolomite. Evidence for replacement are the euhedral rhombs of dolomite surrounded by calcite found in some limestones and dolostones. Replacement occurs after the limestone sediment has been deposited, and the older the rock, the greater the likelihood that the calcite will have been replaced. The source of magnesium has long been a problem to the replacement theory, but one current hypothesis proposes that dolomite may be derived as a byproduct of a reaction that converts montmorillonite to illite in mudrocks. Another means by which dolomite may replace limestone is evaporative pumping of calcium sulfate brines through tidal flat deposits. Evaporative pumping increases the Mg/Ca ratio, and thus, Mg ions replace Ca ions. Groundwater brine replacement of Ca by Mg is another competing hypothesis.

10.4 THE EVAPORITES

Evaporites are sedimentary rocks composed of minerals deposited by evaporation from natural solutions. Two types of evaporites are recognized: marine and continental.

Marine Evaporites

As the term implies, *marine evaporites* are formed from seawater evaporation. These evaporite deposits have a predictable mineral assemblage consisting mostly of gypsum, anhydrite, halite, sylvite, carnallite, and epsomite. However, the relative amounts of these minerals in evaporite sequences varies greatly, and in many instances the minerals resulting from complete evaporation may not be present in the sequence. Evaporite deposits are apparently nearly always produced by partial evaporation. Evaporite deposits are found in the United States Gulf Coast where Jurassic age salt forms a large number of salt domes, as well as in Germany, in the interior basins of the Mid-continent, and the Rocky Mountain region of North America.

The Permian Basin In the Permian Basin of west Texas and eastern New Mexico evaporite deposits more than 3,500 feet thick occur within a few hundred feet of the surface. The Salado Formation, is at 2,000 ft thick, the dominant unit in the sequence, and consists of nearly pure halite. How much seawater would have been needed to precipitate this volume of halite? Because seawater contains about 3.5 ounces of salt per gallon of water, it would have required an ocean almost 100 miles deep *if* it had completely evaporated. This is an entirely unreasonable number, especially since this is a small area on the continent and is unlikely to have ever reached depths of more than a few hundred feet during the time of halite deposition. The only possible explanation is that seawater was recycled through this area; as it evaporated and deposited its salts, more seawater recycled. This reflux process, operating for a long period of time, and accompanied by subsidence accommodating the increasingly thick deposits would allow such a great thickness to accumulate. The level of salt concentration during this recycling determined the relative abundance of each evaporite rock type.

Continental Evaporites

Continental evaporites produced in arid climates by lake evaporation show much greater mineralogical and chemical variation. The composition of these deposits depends on that of the weathering products of the streams feeding the lake. Halite is a common constituent, such as in the Great Salt Lake of northern Utah, but other lakes are rich sources of boron, e.g., Searles Lake, California, or the soda ash Eocene Green River Lake deposits of Wyoming. The Green River Formation contains a large reserve of shale oil in addition to a large reserve of sodium carbonate minerals, and is an important source rock for Utah oil fields (Tuttle, 1991).

10.5 OTHER CHEMICAL AND ORGANIC SEDIMENTS

A number of other precipitates or organic accumulations form bedded deposits. These include bedded cherts and novaculite, coal, phosphorite, banded-iron formations (BIF), and ironstone.

Bedded cherts occur in sequences of sedimentary rocks worldwide. *Novaculite* is a light-colored, extremely fine-grained variety of chert prominent in the Ouachita foldbelt of Arkansas and eastern Oklahoma, and the Marathon Basin of west Texas. The former is of Ordovician, and latter of Devonian, age. The shallow versus deep water origin of these siliceous beds has been an object of intense discussion for decades.

Coal is considered a sedimentary rock by most geologists because it occurs in layers or seams and contains up to 50 percent clay minerals and/or silt. It results from rapid deposition of plant material accumulating in an oxygen-poor environment that inhibited decay, forming a material called peat. The peat subsequently changed to coal by pressure and dewatering accompanying burial. Most coal deposits formed in deltaic environments where subsidence and burial was rapid. Sandstone and shale are commonly interbedded with coal seams. Evidence for coal origins is found in the abundant impressions of leaves occurring in coal seams. A number of coal types are recognized, ranging from *lignite*, the lowest grade coal, to *bituminous* coal, the most abundant type, to *anthracite*, the highest (metamorphic) grade coal. The change in grade reflects the temperature levels reached following burial: the higher the temperature, the higher the grade of coal produced. With increasing grade, its heat energy value (BTU) increases, and ash and sulfur percentages decrease.

Phosphorite is a sedimentary rock rich in phosphate (more than 20 percent P_2O_5). Although phosphorite beds occur in a normal stratigraphic sequence of marine sediments, the formation of large phosphorite deposits requires special circumstances. The continental shelf edge where deeper water rich in phosphate upwells is necessary for its formation. Examples of phosphorite deposits are the widespread Miocene Epoch phosphates of Florida and the Carolinas and the Permian Period Phosphoria Formation of the northern Rocky Mountains which formed on the former west edge of the North American shelf. The Phosphoria Formation is the best studied marine phosphorite (McKelvey et al., 1959). In its type area of southeastern Idaho, it consists of chert, phosphatic and carbonaceous mudstone, phosphorite, cherty mudstone, and minor amounts of dark carbonate rock (Figure 10.6).

Iron-rich sedimentary rocks occur in two different settings and age groupings: the Precambrian *banded-iron formations* (BIF) and the oölitic ironstones, most commonly of Paleozoic age.

Precambrian BIF consist of alternating bands of magnetite and/or hematite, and some form of silica. Because some have undergone extensive metamorphism, a

FIGURE 10.6

Stratigraphy of phosphatic units, Permian age, Idaho and Wyoming.
Source: McKelvey et al. (1959). "The Phosphoria, Park City, and Shedhorn formations in the western phosphate field." *U.S. Geol. Surv. Prof. Pap.* 313-A, 47 pp.

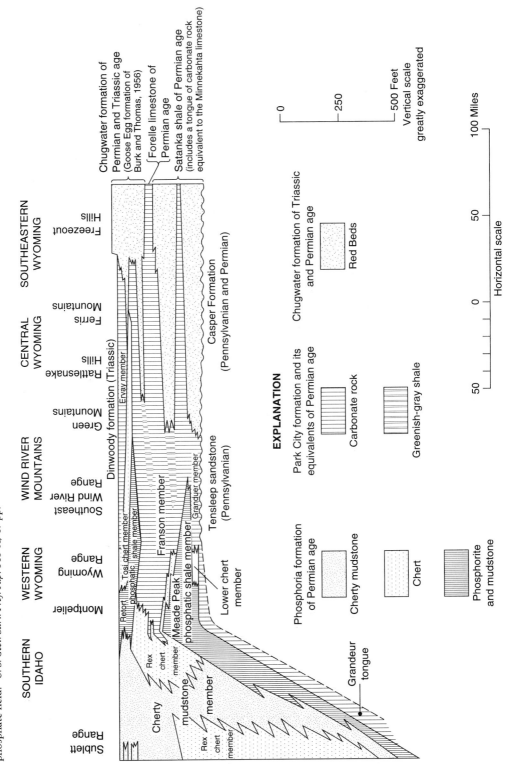

FIGURE 10.7
A septarian nodule showing septa (partitions) giving this feature its name. Septa formed when original sediment dried up, cracked, and new material was deposited in fractures froming partitions. Nodule, 25 cm in longest dimension.

wide range of minerals has been reported from these units. No current analog of these units can form because their origin was due, in part, to the lack of atmospheric and dissolved oxygen. Most BIF formed before the oxygen level reached 1 to 2 percent of its present volume.

The oölitic iron beds look much like carbonate oölite beds except that they are composed of hematite instead of calcite. The most widely accepted hypothesis of origin is that they originated as carbonate layers and subsequently were replaced by iron oxide. Low-temperature geochemists have recently proposed that they could be diagenetic laterite soils. The best known example of oölitic ironstone is the Silurian Clinton Formation exposed in the Appalachian basin of the eastern United States.

10.6 CONCRETIONS, NODULES, AND ARMORED MUDBALLS

Concretions and *nodules* are chemical segregations that occur in clastic and chemical sedimentary rocks, but differ in composition and structure from the host rock. Concretions of pyrite and marcasite are common in black shales where their high specific gravity arouses curiosity. Calcite concretions are common in a variety of rock types. Nodules of chert are common in carbonate rocks, but the most spectacular form are *septarian nodules* or concretions (Figure 10.7). The similar *ironstone concretions* famed for their enclosed plant fossils are formed in clay sediment where organic matter attracts Fe, S, and Ca, producing concentrically layered, hard, spherical to flattened ellipsoidal nodules.

Armored mudball is a rather fearsome sounding name for a spherical mud or silty accumulation armored by granules or pebbles. These features form when the mud is still plastic and is ripped up from the bottom of a streambed. As it moves down stream, coarser particles stick to the outside of the mudball, producing its "armor" (Figure 10.8).

10.7 CLASSIFICATION OF CHEMICAL AND ORGANIC SEDIMENTARY ROCKS

Giving a name to a chemical sedimentary rock can be a relatively simple exercise. However, carbonate rocks have a profusion of names applied to them. Even here a little common sense suggests that perfectly adequate, easily understood names can be applied. Carbonate rocks, like all of the rock types described, have spawned a variety of classification schemes (Folk, 1959, 1974; Dunham, 1962; Leighton & Pendexter, 1962; Mount, 1985). Fine-grained calcite (also known as *micrite*) and coarse-grained or coarsely crystalline calcite (also known as *sparite*) are the two textural

FIGURE 10.8
Armored mudball composed of silt-sized interior, "armored" by granules and pebbles. Mudball is 3 cm in diam.

varians of limestone (Folk, 1974). Dolomite equivalents are dolomicrite and dolosparite. Although dolomite is typically more brown than calcite, the student can readily and quickly test for calcite and dolomite with a drop or two of dilute hydrochloric acid. Calcite effervesces in dilute hydrochloric acid, but dolomite effervesces only if powdered. The essential elements of any classification scheme are mineralogy, texture, and the presence or absence of fossils (Table 10.1).

A fossil-bearing limestone or dolostone unit should be noted as "fossiliferous." A typically brief but complete rock name for a nonfossiliferous carbonate would be "light gray, fine-grained limestone." Another unit may be described as "brownish-yellow, fine-grained, fossiliferous dolostone." Some carbonate rocks are composed almost entirely of cemented shell remains. Such rocks are called *coquina* (Figure 10.9). Another example of a good name would be "light brownish gray, fine-grained fossiliferous limestone" (Figure 10.10). Evaporites can be named using a similar system, but utilizing the dominant mineral as the root rock name (Warren, 1989); for example, "very light gray, fine- to medium-grained, laminated gypsum" (Figure 9.3). The general recipe for a viable name is: 1) color, 2) grain size, 3) fossils (if present), 4) structures or textures (if present), and 5) root lithologic name.

FIGURE 10.9
Coquina composed of cemented oyster shells. From Del Rio Clay, central and west Texas.

FIGURE 10.10
Fossiliferous limestone composed of bivalves and gastropods fragmented, Pennsylvanian age, Sandia Mountains, New Mexico.

11
Metamorphism and Metamorphic Rocks

11.1 INTRODUCTION

So far in the journey of discovery through the mineral and rock world we have progressed from the simplest to the more complex. The *metamorphic* rocks (*meta* = change; *morph* = form) have, in general, more complex histories than either igneous or sedimentary rocks. As the name implies, metamorphic rocks are *changed* rocks. This change, or *metamorphism*, usually occurs by an increase in temperature, pressure, or both. An accompanying change in chemical composition, *metasomatism*, may also occur. All of these changes occur underground out of direct observation. Determining the changes that took place is a process requiring an understanding of metamorphism's effects on the original, unmetamorphosed rock. The two main objectives of the study of a metamorphic rock are: 1) determining the temperature and pressure at which the change (metamorphism) occurred, and 2) determining the original, unmetamorphosed, rock type or *protolith*. The data needed to accomplish these objectives are the rock's mineral assemblages and the presence, if any, of inherited or *relict textures*. Bulk chemical composition can also be used unless there is evidence that this composition changed during metamorphism.

11.2 TYPES OF METAMORPHISM

The two major types of environments in which metamorphism occurs are regional metamorphism, and thermal, or contact, metamorphism.

Regional Metamorphism

Regional metamorphism occurs in orogenic belts and over large areas, generally more than 1,000 square miles. Regionally metamorphosed rocks are *foliated* because of the presence of a directed, tectonic pressure that forces the micaceous minerals, or in some instances prismatic minerals such as hornblende, to grow in preferred planar orientations producing a planar fabric, or foliation, in the rock (Figure 11.1). Metamorphism results from an increase in both temperature and pressure (Figure 11.2).

Rocks produced by regional metamorphism are most common and abundant. Minerals produced by regional metamorphism include garnet, staurolite, cordierite, hornblende, and all of the Al_2SiO_5 minerals, as well as several varieties of mica and feldspar. Products of this metamorphism type include the foliated rocks—schists, gneisses, and phyllites, as well as the amphibole-rich rocks called *amphibolites*. In addition, *migmatite*, a hybrid rock composed in part of melted rock and of metamorphic material, occurs in some regional metamorphic terranes. These migmatites generally consist of granitic veins apparently interbedded or interleaved with foliated mafic metamorphic material. The formation temperature of these hybrid igneous/metamorphic rocks straddles the boundary between igneous and metamor-

FIGURE 11.1
Foliated regionally metamorphosed rock of Precambrain age from roadcut, central Colorado. Photograph shows area approx. 30 cm wide.

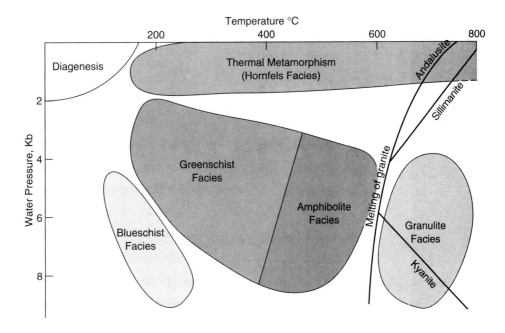

FIGURE 11.2
Temperature and pressure of metamorphism.
Source: Althaus, E. (1967). *Contr. Mineral. Petrol.* v. 13.

phic conditions. They may form by injection of magma into metamorphic rock, by limited in-place melting, or a combination of these processes. As the temperature increases to the lowest melting point of the crust, feldspar and quartz start to melt, forming veins and layers. However, the more refractory minerals, such as biotite, hornblende, and other mafic minerals tend not to melt, and thus, retain their metamorphic texture. The result is a migmatite.

The *granulite facies*, on the right-hand side of the diagram in Figure 11.2, lies in an area seemingly out of metamorphism's realm. However, under certain high pressure and temperature conditions, granulites can form. *Granulites* are generally medium-grained rock composed of feldspar (plus or minus quartz) with hypersthene, or clinopyroxene, almandine, and quartz (Winkler, 1976). It may also contain biotite, hornblende, and sillimanite, or kyanite. Most typical sedimentary or igneous rocks would melt at the temperatures of the granulite facies, producing migmatite or even granite. But under conditions of very little water content, so that water pressure (P_{H_2O}) is much less than the total pressure (P_{Total}), metamorphism will occur rather than melting. The unanswered question is: How is the P_{H_2O} lowered below that of P_{TOTAL}? There are several possibilities:

1. Water may have been lost during previous episodes of metamorphism.
2. Metamorphism may occur only in the deep crust where water is rare.

3. The fluid phase is diluted by another phase such as CO_2. Some evidence for dilution by CO_2 comes from fluid inclusion studies. It is likely that all of the possibilities listed, as well as others, contribute to granulite formation.

Granulites are known to occur only in deep-seated, high-grade metamorphic rocks usually appearing to have undergone multiple metamorphic episodes.

Regional metamorphic terranes occur over large continental areas known as *Precambrian shields* (exposed cratonic rocks). Most Precambrian rocks are igneous and metamorphic. In North America, the Canadian Shield (exposed over the eastern two-thirds of Canada) contains several provinces of igneous and metamorphic rocks ranging in age from about 1 billion to 3.96 billion years. Precambrian igneous and metamorphic rocks also underlie most of North America beneath the veneer of Phanerozoic sedimentary rocks (the platform) covering most of the continent south of the Canadian Shield. The cores of most mountain uplifts in the Rocky Mountains also expose Precambrian regionally metamorphosed rocks much older than the mountains. These rocks record mountain range formation more than a billion years before the birth of the present Rocky Mountains.

Thermal Metamorphism

Thermal or *contact metamorphism* occurs over small areas adjacent to an igneous intrusion (Figure 11.3). The size of the area affected depends on the intrusion's size and temperature. Such rocks do not exhibit foliation, but they may be banded and they are likely to show the effects of a changed chemical composition. The general name for a fine-grained rock not exhibiting foliation or lineation, and resulting from contact metamorphism, is *hornfels*. Fluids derived from adjacent igneous intrusions commonly invade the country rock, adding a variety of elements to the system. The mineralogical and textural changes produced are due mostly to increased temperature. Minerals typical of contact metamorphic deposits include vesuvianite, garnet,

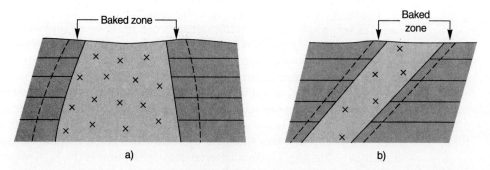

FIGURE 11.3
Environments of contact metamorphism: a) adjacent to a pluton, and b) adjacent to a dike.

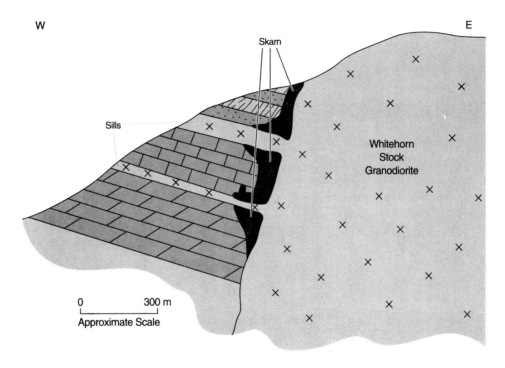

FIGURE 11.4
Cross-section of Whitehorn stock and adjacent Calumet skarn deposit, Chaffee County, Colorado. Skarn deposits shown in black.

corundum, pyroxene, epidote, scapolite, and wollastonite, although several of these minerals may also occur in regionally metamorphosed deposits.

Skarns are another common contact metamorphic deposit. A *skarn* is a metamorphic assemblage of calcium and magnesium silicates, usually with some added Fe. They usually form by contact metamorphism of limestone or dolostone. Many skarns are zoned by element diffusion across the igneous/wallrock contact (Shay, 1975) and most are relatively coarse grained. A good example is in the Calumet District, Chaffee County, Colorado (Figure 11.4). This deposit occurs on the west side of the Whitehorn Stock, a late Cretaceous age granodiorite intrusion. The stock intruded a sequence consisting of lower dolostones, a middle limestone unit, and an upper sequence of sandstones and interbedded shales. Excellent skarn development occurred in the Mississippian Period Leadville Limestone, a nearly pure limestone in the middle of the sequence. The mineral assemblages produced include epidote, garnet, magnetite, actinolite, quartz, and diopside (Figure 11.5). The abundant iron minerals, as well as abundant silicates typical of a skarn, indicate a contribution of Fe and SiO_2 from the stock.

FIGURE 11.5
Banded skarn, Calumet District. Light-colored bands are composed of radiating, prismatic crystals of actinolite, dark bands consist of magnetite. Width of photograph, 10 cm.

Other Types of Metamorphism

Other types of metamorphism include very low grade metamorphism sometimes called burial or zeolite facies metamorphism, and dynamic metamorphism. *Very low grade metamorphism* (Winkler, 1976) is characterized by low temperatures, ranging from 175°C to 400 °C, and pressures ranging to as much as that of 40-km burial depths. At lower pressures, in the realm of burial metamorphism, a mineral assemblage characterized by the minerals heulandite, laumontite, and pumpellyite is produced. At higher pressures, ranging from six to nine kilobars, equivalent to a depth of 25 to 40 km, the characteristic mineral assemblage includes glaucophane, lawsonite, jadeite, and aegirine. These minerals produce a distinctly blue-colored schist giving rise to the common name for this assemblage—*blueschist*. Blueschists occur in metamorphic belts (Miyashiro, 1961) and are particularly well developed around the Pacific Ocean margin parallel to coastlines. Good examples are found on the western margin of North America and in the Japanese archipelago. In North America, isolated blocks of blueschist and amphibolite facies occur in Jurassic and Cretaceous period deposits in the Coast Range of Oregon and California (Coleman & Lanphere, 1971).

Eclogites are fine- to coarse-grained rocks composed of dark green omphacite speckled with reddish-brown garnet porphyroblasts up to one cm in diameter. Omphacite is a Na-Ca pyroxene, and omphacite and garnet typically make up more than half, and as much as 90 percent of the rock. Minor minerals include rutile, glaucophane, hornblende, and sphene. Eclogites have a basaltic or gabbroic chemical composition. Because they lack plagioclase, an essential constituent of basalts, they are thought to have formed at very high pressures (>10 Kb), resulting in the disappearance of plagioclase.

Eclogite usually occurs as blocks or lenses in blueschist outcrops, and as nodules or inclusions in kimberlites and basalts (Winkler, 1976). It is common in the Franciscan Formation of northern California (Hyndman, 1972) where it occurs as rounded blocks or boulders from a few feet to several hundred feet long.

Dynamic metamorphism, also called dislocation metamorphism, occurs along *shear zones* in the crust. The shearing stresses result in mechanical grinding of rock accompanied by recrystallization (growth of crystals to larger sizes). The net effect is a *cataclastic texture* characterized by fragments of pre-existing crystals, usually somewhat rounded, and aligned in layers in a groundmass (fine-grained matrix) of powdered, granulated rock. Larger amounts of shearing reduce the large crystal fragments' sizes, producing a finer-grained, banded rock called *mylonite*. With increasing shearing stress and intensity of deformation, the grain size decreases and the degree of recrystallization increases, producing an *ultramylonite*.

11.3 PRESSURES AND TEMPERATURES OF METAMORPHISM

The conditions of metamorphism can be illustrated on a pressure-temperature diagram depicting the cessation of diagenesis on the low-temperature end, and the beginning of melting at the high-temperature end (Figure 11.2).

Regions within this diagram are delineated as metamorphic facies. The term, *facies*, is an old one that has been applied in a variety of ways. In metamorphic petrology, facies designates a group of rocks metamorphosed at approximately the same temperature and pressure conditions.

Contact metamorphosed rocks, also known as the *hornfels facies*, metamorphose under a wide range of temperatures, but at low pressure, and they plot along the diagram's upper margin. All temperatures less than about 175 °C lie within the realm of weathering, erosion, or diagenesis. Very low-grade metamorphism products are represented on the diagram's left margin as the *blueschist facies*. The diagram's central portion represents *greenschist* and *amphibolite facies* of regional metamorphism. The greenschist facies' lower temperature boundary is marked by the disappearance of pumpellyite or lawsonite, and the formation of zoisite or clinozoisite. Other typical minerals of the greenschist facies include chlorite, actinolite, albite, epidote, and muscovite. The lower boundary of the amphibolite facies is marked by the appearance of cordierite or staurolite, and chlorite's disappearance. Other typical minerals include oligoclase or andesine plagioclase, kyanite, sillimanite, and microcline. The common garnet of regional metamorphic rocks, almandine, first appears in the upper part of the greenschist facies and persists into the amphibolite facies. The upper boundary of amphibolite facies metamorphism marks the *granite melting curve*. This curve is the upper limit for typical metamorphic rocks.

The *granulite facies*, on the right, lies in an area seemingly out of metamorphism's realm—except under conditions of very low P_{H_2O}.

11.4 TEXTURES AND STRUCTURES

We have already briefly discussed some of the textural terminology of the metamorphic rocks. A relatively long list of textural terms has been proposed, but a few simple terms suffice for most rocks. Any *inherited texture* in a metamorphic rock is denoted by the prefix *"blasto."* For example, an inherited porphyritic texture is called *blastoporphyritic*. However, a porphyritic texture produced through metamorphism is denoted by the *suffix* "blasto," e.g., *porphyroblastic*. Metamorphic phenocrysts are called *porphyroblasts*, e.g., staurolite porphyroblasts.

Another common texture is *augen*. Derived from the German word for "eye," this is a lenticular mineral grain texture (Figure 11.6).

FIGURE 11.6
Augen texture in gneiss from Wet Mountains, Colorado. Augen consists of alkali feldspar with "pupils" of magnetite (4 to 6 mm in diameter).

Many gneisses exhibit *cataclastic texture*, resulting from mechanical fragmentation. Typical of this texture are lens-like fragments of porphyroblasts in a fine-grained matrix of crushed mineral grains.

11.5 CLASSIFICATION OF METAMORPHIC ROCKS

The classification of metamorphic rocks is a reasonably simple procedure that begins by studying texture. Determining the presence or absence of foliation is the first step. Foliation in many rocks is expressed as *schistocity*, a parallel orientation of mica grains. Banding may also be present, but this alone does not mean the rock is foliated (Figure 11.5). For example, banded skarns are common, but distinguished from foliated regionally metamorphosed rocks by their arrangement of minerals in nonparallel bands. The prismatic crystals of actinolite shown in Figure 11.5 do not parallel the banding. Table 11.1 gives the general scheme for applying root names to metamorphic rocks and determining probable protolith.

Foliated Rocks

Foliated rocks include—in order of increasing grain size—slate, phyllite, schist, and gneiss. The latter two rock types have comparable grain sizes, but schists are characterized by abundant mica, whereas gneisses are richer in feldspar and quartz. The foliation in each is expressed in different ways. *Slate* superficially resembles shale but is much harder, and breaks into thin, planar fragments parallel to the foliation. This parting or *cleavage* is so perfect that slate is used in the manufacture of high-quality billiard tables and was formerly used to make chalkboards. In the New England area, many buildings are roofed with slate shingles. *Phyllite* is slightly coarser grained but the individual mica flakes are not visible to the unaided eye. However, parallel arrangement of the mica produces a silky sheen or luster on surfaces parallel to the foliation. Another characteristic of the phyllites is the *crenulations*, forming "wrinkles" on the foliation surfaces. Grain size in schists and gneisses may range from the barely visible to mica flakes more than an inch in diameter. Where mica flakes and more equidimensional minerals such as garnet or quartz occur in the same rock, the mica flakes are typically deformed around the harder mineral grains. Gneisses may not exhibit layering, which is generally visible as banding. The banding is produced by distinctly segregated layers of dark- and light-colored minerals. These layers may have segregated due to metamorphism.

Metamorphic rock root names can be modified in a number of ways to produce a name giving an accurate picture of its appearance. Possible modifiers include igneous rock names, textural terms, and mineral names. Several examples illustrate some of the different techniques for applying such names:

> *Example 1*—A foliated rock composed mostly of visible muscovite flakes with subordinate quartz and feldspar: a brief but complete rock name for this rock is muscovite schist. An acceptable, alternative name is quartz-feldspar-muscovite schist.

Example 2—A foliated rock composed of chlorite and biotite flakes up to 1/2 inch in diameter with garnet porphyroblasts up to 3/4 inch in diameter: understanding that the presence of mica is implied, it may be called a garnet schist. Alternatively it may be called a garnet-biotite-chlorite schist (Figure 11.7).

TABLE 11.1
Classification of Metamorphic Rocks

Fabric	Mineralogy	Possible Protoliths	Rock Name
Nonfoliated	many possible minerals	shale or mudstone	*Hornfels* (fine-grained)
	wollastonite, epidote and other Ca- or Mg-silicates	limestone or dolostone	*Skarn* (coarse-grained)
	calcite or dolomite	limestone or dolostone	*Marble*
	quartz	quartz arenite	*Quartzite*
Foliated	hornblende and plagioclase	gabbro or basalt	*Amphibolite* (medium to coarse-grained)
	not identifiable in hand specimen	shale	*Slate* (very fine-grained)
	micas, garnet, or graphite	shale or claystone	*Phyllite* (fine-grained, pearly luster)
	micas, garnet, staurolite, or cordierite	shale or mudstone	*Schist*
	quartz, feldspar, micas, garnet	granitic igneous rock, arkose	*Gneiss*
	hypersthene, kyanite, or sillimanite	igneous rocks	*Granulite* (medium to coarse-grained)
Cataclastic	quartz and feldspar	granitic igneous rocks	*Mylonite*

Example 3—A gneiss composed of approximately 15 percent biotite, 25 percent quartz, 25 percent plagioclase, and 35 percent alkali feldspar: having a composition of a granite, it may be called a biotite-granite-gneiss, or if it retains evidence of an original igneous origin, a granitic biotite-gneiss. In the latter case, the igneous term describes composition and the metamorphic term describes texture. Other examples of this type of hybrid naming scheme include names such as diorite gneiss or granodiorite gneiss.

Nonfoliated Rocks

Naming nonfoliated rocks is a slightly more difficult task. Some of these rocks are fine grained and thus, determining their mineral composition can be problematic in hand-specimen. The general term for a fine-grained rock not exhibiting foliation is *hornfels*. Some nonfoliated rocks may result from regional metamorphism, yet may not contain minerals producing foliation. Examples of such rocks include marble and quartzite, and in some instances, amphibolite. Some amphibolites may show foliation and even *lineation* (Figure 11.8). Recall that skarns may be banded or nonbanded. The root name modifers are similar to those of foliated rocks. A few examples (page 185) will serve to illustrate the general procedure.

FIGURE 11.7
Garnet-chlorite-biotite schist. Garnet porphyroblasts are approximately 1 cm in diameter.

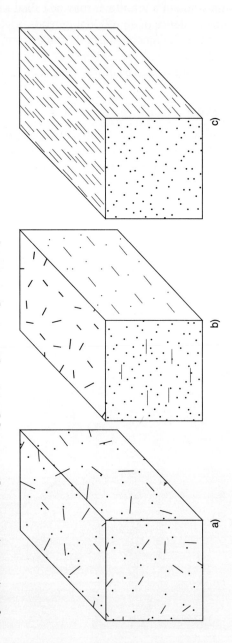

FIGURE 11.8
Block diagrams showing how prismatic mineral produces foliation or lineation: a) no foliation or lineation present; b) foliation produced by linear elements (prismatic hornblende crystals); and, c) lineation produced by parallel alignment of prismatic crystals.

Example 1—A fine-grained rock not exhibiting foliation, with a few dodecahedral crystals of garnet visible in the hand-specimen: the name would be garnet hornfels.

Example 2—A coarsely crystalline rock composed of calcite (effervesces in dilute HCl) with thin prismatic crystals of wollastonite and green, blocky crystals of vesuvianite present in smaller amounts: this may be known as wollastonite-vesuvianite marble. It may also be called a wollastonite-vesuvianite skarn if the silicate minerals make up a major portion of the rock.

Rock description and identification is not always easy, but the student should be willing to take the plunge. Remember that any rock name can always be changed. Practice makes perfect and the satisfaction of placing a proper name on a rock cannot be underestimated.

APPENDIX I
The Geologic Time Scale

FIGURE AP1-01

The Geologic Time Scale. (Numbers at boundaries between eras represent time in millions of years before present.)

Source: Palmer, A.R. (compiler) (1983). The Decade of North American Geology. Geologic Time Scale. *Geology*, v. 11, p. 503–504.

APPENDIX II
Elements of Crystallography

Crystallography is a subdiscipline of mineralogy concerning crystal geometry and the study of mineral symmetry and crystal form. This crystal form or morphology is the natural result of the mineral's internal ordered structure. However, variations in the growth environment can change crystal morphology. For example, no two quartz crystals are exactly alike, even though all have the same internal arrangement of Si and O atoms. The differences among quartz crystals arise from conditions of growth, available space for growth, and growth imperfections. However, although no two crystals are identical, the angle between adjacent crystal faces remains constant. This constancy of the "interfacial angle" was noted by Steno in the 17th century. So crystal form or morphology is a valuable and useful tool for identification when crystals are present.

Mineralogists classify minerals into one of six different crystal systems. In order from highest to lowest symmetry, these systems are isometric, hexagonal, tetragonal, orthorhombic, monoclinic, and triclinic. Higher symmetry crystals are more symmetrical, with many mirror planes—imaginary planes dividing them into two mirror-images. A mirror plane is one of several symmetry elements. The larger the number of symmetry elements, the higher the crystal's symmetry. Other symmetry elements include a center of symmetry and axes of rotation. For example, a mineral has a four-fold axis of rotation if it can be rotated about an axis, repeating itself four times in one revolution. The typical crystal forms for each crystal system are given in the following list.

Isometric Crystals in this system are generally equidimensional, having four, six, eight, or 12 faces. Examples of crystal forms in this system are the cube, ocatahedron (eight faces), and dodecahedron (12 faces).

Hexagonal This crystal system has two common forms. Some crystals have a hexagonal prism (six crystal faces parallel to one axis), and may be capped at either end with a hexagonal pyramid (six faces meeting at a point). The other common form in this system is the rhombohedron, which resembles a deformed cube.

Tetragonal The common forms in this system are tetragonal (four-sided) prisms and pyramids. Most forms possess two or more mirror planes.

Orthorhombic Crystals in this system are very similar to those of the tetragonal system. Prisms capped by pyramids, many composed of a large number of faces, are typical of this system. Most forms possess two or more mirror planes.

Monoclinic and Triclinic These systems comprise similar crystals of low symmetry. Interfacial angles are not orthogonal, and many appear skewed or deformed. Monoclinic crystals usually possess one mirror plane, and triclinic crystals none.

Appendix II

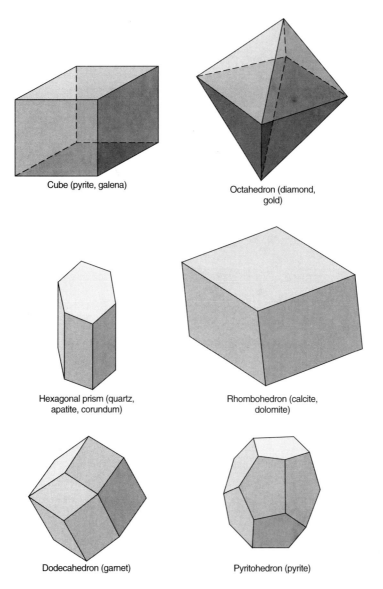

FIGURE AP2-01
Crystal forms and typical minerals that exhibit these forms

APPENDIX III
Bowen's Reaction Series

FIGURE AP3-01
Bowen's Reaction Series describes the order of precipitation of minerals from a basaltic magma. High-temperature minerals crystallize first, low-temperature minerals crystallize last.

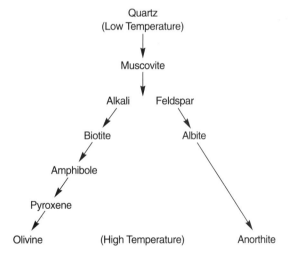

Because the chemical composition of the minerals is different from the composition of the melt (initially a basaltic magma), the crystallization of the magma changes the composition of the remaining melt, by depleting the magma in the elements incorporated in the crystallizing minerals. Also, the system has the potential to produce a variety of rock types if the minerals that crystallize are separated from the remaining melt, for example, by gravitational settling. Thus an early-formed mineral suite of olivine and pyroxene would yield a peridotite rock, an intermediate plagioclase and hornblende suite would form a gabbro, and a late-forming suite of quartz and feldspar would yield a granitic rock.

APPENDIX IV
Table of Elements

Element	Symbol	Atomic Number	Atomic Weight	Element	Symbol	Atomic Number	Atomic Weight
Actinium	Ac	89	[227]	Erbium	Er	68	167.2
Aluminum	Al	13	26.9	Europium	Eu	63	151.9
Americium	Am	95	[243]	Fermium	Fm	100	[253]
Antimony	Sb	51	121.7	Fluorine	F	9	18.9
Argon	Ar	18	39.9	Francium	Fr	87	[223]
Arsenic	As	33	74.9	Gadolinium	Gd	64	157.25
Astatine	At	85	[210]	Gallium	Ga	31	69.72
Barium	Ba	56	137.3	Germanium	Ge	32	72.59
Berkelium	Bk	97	[249]	Gold	Au	79	196.9
Beryllium	Be	4	9.0	Hafnium	Hf	72	178.49
Bismuth	Bi	83	208.9	Helium	He	2	4.0
Boron	B	5	10.8	Holmium	Ho	67	164.9
Bromine	Br	35	79.9	Hydrogen	H	1	1.0
Cadmium	Cd	48	112.4	Indium	In	49	114.82
Calcium	Ca	20	40.0	Iodine	I	53	126.9
Californium	Cf	98	[251]	Iridium	Ir	77	192.2
Carbon	C	6	12.01	Iron	Fe	263	55.8
Cerium	Ce	58	140.12	Krypton	Kr	6	83.80
Cesium	Cs	55	132.9	Kurchatorium	Ku	104	[260]
Chlorine	Cl	17	35.4	Lanthanum	La	57	138
Chromium	Cr	24	51.9	Lawrencium	Lw	103	[257]
Cobalt	Co	27	58.9	Lead	Pb	82	207.19
Copper	Cu	29	63.5	Lithium	Li	3	6.93
Curium	Cm	96	[247]	Lutetium	Lu	71	174.97
Dysprosium	Dy	66	162.5	Magnesium	Mg	12	24.3
Einsteinium	Es	99	[254]	Manganese	Mn	25	54.9

FIGURE AP4-01
Table of the elements showing their symbols, atomic numbers, and atomic weights.

Appendix IV

Element	Symbol	Atomic Number	Atomic Weight	Element	Symbol	Atomic Number	Atomic Weight
Mendelevium	Md	101	[256]	Ruthenium	Ru	44	101.07
Mercury	Hg	80	200.5	Samarium	Sm	62	150.35
Molybdenum	Mo	42	95.9	Scandium	Sc	21	44.9
Neodymium	Nd	60	144.2	Selenium	Se	34	78.96
Neon	Ne	10	20.18	Silicon	Si	14	28.086
Neptunium	Np	93	[237]	Silver	Ag	47	107.870
Nickel	Ni	28	58.71	Sodium	Na	11	22.9
Niobium	Nb	41	92.9	Strontium	Sr	38	87.62
Nitrogen	N	7	14.0	Sulfur	S	16	32.0
Nobelium	No	102	[254]	Tantalum	Ta	73	180.9
Osmium	Os	76	190.2	Technetium	Tc	43	[99]
Oxygen	O	8	15.9	Tellurium	Te	52	127.60
Palladium	Pd	46	106.4	Terbium	Tb	65	158.9
Phosphorus	P	15	30.97	Thallium	Tl	81	204.37
Platinum	Pt	78	195.09	Thorium	Th	90	232.0
Plutonium	Pu	94	[242]	Thulium	Tm	69	168.9
Polonium	Po	84	[210]	Tin	Sn	50	118.69
Potassium	K	19	39.102	Titanium	Ti	22	47.90
Praseodymium	Pr	59	140.907	Tungsten	W	74	183.85
Promethium	Pm	61	[147]	Uranium	U	92	238.03
Protactinium	Pa	91	[231]	Vanadium	V	23	50.9
Radium	Ra	88	[226]	Xenon	Xe	54	131.30
Radon	Rn	86	[222]	Ytterbium	Yb	70	173.04
Rhenium	Re	75	186.2	Yttrium	Y	39	88.9
Rhodium	Rh	45	102.9	Zinc	Zn	30	65.37
Rubidium	Rb	37	85.47	Zirconium	Zr	40	91.22

FIGURE AP4-01
Continued

Glossary

Aa A textural type of lava characterized by a blocky surface.

Accessory A mineral present in minor amounts in a rock and not one of the essential minerals.

Anion A negatively charged ion, produced by a gain of electrons.

Ash Flow A turbulent volcanic flow that moves down the slopes of a volcano at high speeds. Although composed of ash (glass shards, dust, crystals), it behaves like a lava flow.

Asterism A property of some minerals that produces a star-shaped reflection.

Asthenosphere The partially molten portion of the upper mantle that underlies the lithosphere. It extends from a depth of about 100 km to a depth of about 700 km.

Atom The smallest part of an element that retains all of the characteristics of that element. It is composed of protons, electrons, and neutrons.

Authigenic In a sedimentary rock, refers to material formed in place.

Batholith A large intrusive body or a combination of several smaller intrusions (composite batholith).

Body Waves Earthquake waves that traverse the interior of the Earth. Includes the Primary (P) and Secondary (S) waves.

Bulbous Budding The process by which pillow lavas are formed. Takes place beneath the surface of a body of water.

Caldera A generally circular collapse zone formed at the summit of volcanoes (summit caldera) or by the eruption of voluminous amounts of ash-flow tuff (collapse calderas).

Cation   A positively charged ion, formed by a loss of electrons.

Chemical Sedimentary Rock   A type of sedimentary rock formed by precipitation from fresh or oceanic water.

Chill Zone   The fine-grained margin of an igneous intrusive rock where the cooler wall rock promoted rapid cooling of the magma.

Cinder Cone   A volcanic structure composed of cinders that accumulate around the vent.

Clastic   A type of sedimentary rock composed of fragments of older rocks.

Cleavage   In a mineral, the tendency of the mineral to break along smooth plane reflective surfaces.

Color Index   The volume percent of mafic minerals in an igneous rock, which are all minerals except the essential minerals.

Composite Volcano   A volcano composed of lava and cinders.

Diagenesis   Processes that change a sediment after it has been deposited, including compaction, cementation, and recrystallization.

Dike   A tabular, discordant intrusive body.

Dispersion   An optical property of transparent minerals. A measure of the ability of the mineral to break up a beam of white light into its constituent colors.

Eclogite   An ultramafic rock composed of garnet and omphacite.

Electrum   A natural alloy of silver and gold containing at least 20 percent silver.

Endmember   In a solid solution series, one of the compositional extremes.

Eolian   Of or pertaining to the wind.

Epicenter   The location on the Earth's surface directly above the focus.

Essential Mineral   A mineral essential to the classification of a rock. In igneous rocks the essential minerals are quartz, feldspars, and feldspathoids.

Euhedral   Used to describe a crystal bounded by smooth plane crystal faces.

Eutaxitic Texture   The texture produced by flattened pumice fragments in an ash-flow tuff or welded tuff.

Evaporite   A chemical sedimentary rock produced by evaporation of lake or seawater.

Focus   The site of an earthquake below Earth's surface.

Foliation   A property of regionally metamorphosed rocks characterized by a parallel alignment of mica minerals.

Fossil   Evidence of an ancient organism.

Fracture   A physical property of minerals. Typical of minerals that do not exhibit cleavage.

Gangue   Minerals in an ore deposit that have no economic value.

Geologic Time Scale   The subdivision of geologic time into eras and periods. See Appendix I.

Geotherm   A graph of the change in temperature with depth within the Earth.

Graphic Granite   A common rock in pegmatites consisting of an intimate intergrowth of quartz and alkali feldspar producing a cuneiform pattern.

Gutenburg Discontinuity   The boundary between the mantle and the core.

Hardness   A physical property of minerals that measures their resistance to scratching.

Herringbone A texture resembling a feather or fish skeleton. Found in some native metals.

Hornfels A fine-grained rock that is the result of contact metamorphism.

Hydrothermal A class of ore deposits formed by hot fluids below Earth's surface.

Ignimbrite See Welded Tuff.

Intrusive An igneous rock body that has been intruded and cooled beneath Earth's surface.

Ion A charged atom, produced by a loss or gain of electrons.

Isotope A species of a chemical element that contains a different number of neutrons than other species of the same element.

Kamacite A Ni-poor iron alloy found in meteorites.

Laccolith A concordant intrusion with a flat floor and an arched roof.

Liquidus In a phase diagram, the line or surface that defines the lowest temperature at which the system is entirely liquid.

Lithosphere The upper 100-kilometer thick layer of Earth. Includes all of the crust and part of the upper mantle. It is underlain by the asthenosphere.

Luster The surface appearance of a mineral. The two main types are metallic and nonmetallic.

LVZ Low velocity zone, so-called because earthquake waves are observed to slow down when traversing this zone. Equivalent to the asthenosphere.

Magma Molten rock material.

Malleable Can be pounded into shape, i.e., not brittle.

Mesosphere The bulk of Earth. It lies beneath the asthenosphere and includes most of the mantle and all of the core.

Metamorphism Processes that change a rock's mineralogy and texture below Earth's surface due to increased T (Temperature) and/or P (Pressure).

Meteorite Fragments of planetary bodies that originated outside of Earth but within our solar system.

Migmatite A hybrid igneous/metamorphic rock composed of veins of granitic material sandwiched within foliated metamorphic rock.

Mineral A naturally occurring element or compound with a definite chemical composition and an internal ordered structure.

Mode The volume percent of minerals in a rock.

Mohorivicic Discontinuity The boundary between the crust and the mantle.

Mylonite A banded or layered metamorphic rock produced by dynamic (shearing) metamorphism.

Novaculite A light-colored, fine-grained sedimentary rock composed mostly of microcrystalline quartz.

Nucleus The central portion of an atom. Contains the neutrons and protons.

Obsidian A dark volcanic glass, usually rhyolitic in composition.

Octahedrite A type of iron meteorite characterized by Widmanstatten structure.

Oölid Small (less than 2 mm diameter) spherical objects typically found in limestones. The product of calcite precipitaion in a wave or current environment.

Ore Deposit   Any concentration of minerals that can be economically extracted.

Pahoehoe   A textural lava type, usually basaltic, which has a ropy surface texture, and is characteristic of very fluid lava.

Pegmatite   A plutonic igneous rock characterized by very large crystals.

Peridotite   An ultramafic rock consisting of pyroxene and olivine.

Phacolith   A concordant lenticular intrusive body found in the axis of folds.

Phenocryst   The larger crystals in an igneous rock that contains two different size ranges of crystals.

Pillow Lava   A type of lava, usually basaltic, that resembles a stack of pillows, in cross-section. Forms by bulbous budding.

Placer   An ore deposit formed by weathering and accumulation of resistant minerals of above average specific gravity.

Plinian   A vertical volcanic eruption column. Named after Pliny, who described the 79 A.D. plinian eruption of Mt. Vesuvius.

Pluton   A generally cylindrical magmatic intrusion.

Polymorph   Two or more minerals with the same chemical composition but different atomic structure.

Porphyritic   A texture of igneous rocks characterized by crystals of two distinct size ranges. The large crystals are called phenocrysts.

Porphyry Copper   A type of copper ore deposit characterized by porphyritic texture and low grade.

Protolith   In a metamorposed rock, the original, unmetamorphosed rock.

Pyroxenite   An ultramafic rock composed of pyroxene.

Reef   A mining term for an ore-bearing horizon; also a ridge or moundlike structure composed of calcite-secreting organisms such as coral.

Relict Texture   The inherited texture in a metamorphic rock.

Remanant Magnetism   The magnetism preserved in iron-bearing minerals. It records the direction of Earth's magnetic field at the time of their formation.

Rock   An aggregate of one or more minerals.

Sectile   Mineral that can be cut with a knife.

Seismology   The science of the study of earthquakes.

Shield Volcano   A volcano composed of very fluid lava, usually basalt, with a low domal profile reminiscent of a warrior's shield.

Sill   A tabular concordant (parallel to the country rock) intrusive body.

Skarn   A metamorphic assemblage consisting mostly of Ca-Mg silicates.

Solid Solution   Term used to describe a mixture in the solid state. Minerals that exhibit solid solution have a range of compositions, e.g., the feldspars.

Solidus   On a phase diagram, the line or surface that defines the upper temperature boundary of the crystalline (solid) state.

Specific Gravity   The ratio of the density of a mineral or rock to that of an equal volume of water.

Stoichiometric   A mineral whose chemical composition is identical to its chemical formula, e.g., quartz.

Streak   The color of the powdered mineral.

Surface Waves   Earthquake waves that travel along the surface.

Taenite   The Ni-rich iron alloy found in some meteorites.

Ultramafic   A rock rich in iron and magnesium, usually composed of olivine or pyroxene or both.

Viscosity   A measure of resistance to flow.

Wacke   A group of sandstones characterized by a significant amount, more than 20 percent, of a fine-grained matrix.

Welded Tuff   A volcanic rock exhibiting a welded texture of glass shards.

Widmanstatten Pattern   The intergrowth of Ni-rich and Ni-poor iron alloys produced in some iron meteorites such as the octahedrites.

Xenolith   A foreign rock fragment found included within another rock.

References

Althaus, E. 1967. "The triple point andalusite-sillimanite-kyanite." *Contr. Mineral. Petrol.* v. 13, p. 31–50.

Barker, D. S. 1983. *Igneous Rocks*. Englewood Cliffs, NJ: Prentice-Hall. 417 pp.

Berry, L. G., B. Mason, and R. V. Dietrich. 1983. *Mineralogy*. 2nd ed. San Francisco: W. H. Freeman. 561 pp.

Best, M. G. 1982. *Igneous and Metamorphic Petrology*. San Francisco: W. H. Freeman. 630 pp.

Blackburn, W. H., and W. H. Dennen. 1988. *Principles of Mineralogy*. Dubuque, IA: Wm. C. Brown. 413 pp.

Blatt, H. 1982. *Sedimentary Petrology*. San Francisco: W. H. Freeman. 564 pp.

Blatt, H., G. Middleton, and R. Murray. 1980. *Origin of Sedimentary Rocks*. 2nd ed. Englewood Cliffs, NJ: Prentice-Hall. 782 pp.

Boggs, S., Jr. 1987. *Principles of Sedimentology and Stratigraphy*. New York: Merrill/Macmillan. 784 pp.

Brookins, D. G., B. C. Chakoumakos, C. W. Cook, R. C. Ewing, G. P. Landis, and M. E. Register. 1979. "The Harding Pegmatite: Summary of Recent Research." *New Mexico Geol. Society Guidebook, 30th Field Conf.* p. 127–133.

Brown, E. H., B. W. Evans, R. B. Forbes, and P. Misch. 1984. "Blueschists and related eclogites." *Geology*. v. 12, p. 318–319.

Christiansen, R. L., and D. W. Peterson. 1981. Chronology of the 1980 Eruptive Activity. P. W. Lipman and D. R. Mullineaux (Eds). *The 1980 Eruptions of Mount St. Helens, Washington*. U.S. Geological Survey Professional Paper 1250.

Clague, D. A., and G. B. Dalrymple. 1987. "The Hawaiian-Emperor volcanic chain Part I. geologic evolution." In Decker, R. W., T. L. Wright, and P. H. Stauffer (eds). "Volcanism in Hawaii." U.S. Geological Survey Professional Paper 1350, p. 5–54.

Coleman, R. G., and M. A. Lanphere. 1971. "Distribution and age of high-grade blueschists, associated eclogites, and amphibolites from Oregon and California." *Geological Soc. America Bulletin.* v. 82, p. 2397–2412.

Deer, W. A., R. A. Howie, and J. Zussman. *An Introduction to the Rock-Forming Minerals.* 2nd ed. London: Longman Group Limited. 695 pp.

Dietrich, R. V., and B. J. Skinner. 1990. *Gems, Granites, and Gravels.* Cambridge, MA: Cambridge University Press, 173 pp.

Dunham, R. J. 1962. "Classification of carbonate rocks according to depositional textures." In W. E. Ham (Ed.). "Classification of Carbonate Rocks." *Amer. Assoc. Petroleum Geol. Mem.* 1, p. 108–121.

Ehlers, E. G., and H. Blatt. 1982. *Petrology—Igneous, Sedimentary, and Metamorphic.* San Francisco: W. H. Freeman. 729 pp.

Folk, R. L. 1974. *Petrology of Sedimentary Rocks.* Austin: Hemphill Publishing. 182 pp.

Folk, R. L. 1959. "Practical petrographic classification of limestones." *Amer. Assoc. Petroleum Geologists Bull.* v. 43, p. 1–38.

Foshag, W. F., and J. G. R. 1956. "Birth and Development of Paricutin Volcano, Mexico." *U.S. Geological Survey Bulletin* 965–D.

Gruenewaldt, G. Von, M. R. Sharpe, and C. J. Hatton. 1985. "The Bushveld Complex: introduction and review." *Economic Geology*, v. 80, p. 803–812.

Hamilton, W., and W. B. Myers. 1967. "The Nature of Batholiths." U.S. Geological Survey Professional Paper 554–C, 30 p.

Hess, H. H. 1962. "History of ocean basins." In Engel, A. J., H. L. James, and B. F. Leonard (eds). *Petrological Studies: A Volume in Honor of A. F. Buddington. Geological Society of America*, p. 599–620.

Hess, P. C. 1989. *Origins of Igneous Rocks.* Cambridge, MA: Harvard University Press. 336 pp.

Holcomb, R. T. 1976. Preliminary map showing products of eruptions, 1962–1974 from the upper east rift zone of Kilauea Volcano, Hawaii. *U.S. Geol. Survey Miscellaneous Field Studies Map MF-811*, scale 1:24,000.

Hurlbut, C. S. 1970. *Minerals and Man.* New York: Random House. pp. 304.

Hyndman, D. W. 1972. *Petrology of Igneous and Metamorphic Rocks.* New York: McGraw-Hill. 533 pp.

Irvine, T. N., and W. R. A. Baragar. 1971. "A guide to the chemical classification of the common volcanic rocks." *Can. Jour. Earth Sci.*, v. 8, p. 523–548.

Jahns, R. H., and C. W. Burnham. 1969. "Experimental studies of pegmatite genesis: I. A model for the derivation and crystallization of granitic pegmatites." *Economic Geology*, v. 64, p. 843–864.

Jahns, R. H., and R. C. Ewing. 1976. The Harding Mine, Taos County, New Mexico. *New Mexico Geol. Soc. Guidebook, 27th Field Conf.*, p. 263–275.

Jensen, D. E. 1958. *Getting Acquainted with Minerals.* Revised edition. New York: McGraw-Hill. 362 pp.

Jensen, M. L., and Bateman, A. M. 1981. *Economic Mineral Deposits.* 3rd. ed. New York: John Wiley & Sons. 593 pp.

References

Klein, C., and C. S. Hurlbut. 1985. *Manual of Mineralogy*. 20th ed. New York: John Wiley & Sons. 596 pp.

Klein, G. 1963. "Analysis and review of sandstone classifications in the North American geological literature, 1940–1960." *Geol. Soc. America Bulletin*. v. 74, p. 555–576.

Koschman, A. H. 1949. Structural Control of the Gold Deposits of the Cripple Creek District, Teller County, Colorado. *U.S. Geol. Surv. Bull.* 955-B, 57 p.

Lapedes, Daniel N. (ed.). 1978. *McGraw-Hill Encyclopedia of the Geological Sciences*. New York: McGraw-Hill. 915 pp.

Leighton, M. W., and C. Pendexter. 1962. "Carbonate rock types." In Ham, W. E. (Ed.). *Classification of Carbonate Rocks. American Assoc. Petroleum Geol. Mem.* 1, p. 50–69.

LeMaitre, R. W. 1984. A proposal by the IUGS Subcommission on the systematics of igneous rocks for a chemical classification of volcanic rocks based on the total alkali silica (TAS) diagram. *Australian Jour. Earth Sci.*, v. 31, p. 243–255.

Lockwood, J. P., and P. W. Lipman. 1987. "Holocene eruptive history of Mauna Loa Volcano." In Decker, R. W., T. L. Wright, and P. H. Stauffer (eds). *Volcanism in Hawaii*. U.S. Geol. Survey Professional Paper 1350. p. 509–536.

McBirney, A. R. 1984. *Igneous Petrology*. San Francisco: Freeman, Cooper and Co. 504 pp.

McBride, E. F. 1963. "A classification of common sandstones." *Jour. Sed. Petrol.*, v. 33, p. 664–669.

McDonald, D. A., and R. C. Surdam (eds.). 1984. *Clastic Diagenesis*. Tulsa: American Association of Petroleum Geol. 429 pp.

McKelvey, V. E., J. S. Williams, R. P. Sheldon, E. R. Cressman, T. M. Cheney, and R. W. Swanson. 1959. "The Phosphoria, Park City and Shedhorn formations in the western phosphate field." U.S. Geological Survey Professional Paper 313–A, 47 pp.

Meyer, H. O. A. 1977. "Mineralogy of the Upper Mantle: A review of the minerals in mantle xenoliths from kimberlite." *Earth Sci. Rev.*, v. 13, p. 251–281.

Mitchell, R. S. 1973. "Metamict minerals: A review." *The Mineralogical Record*, v. 4, p. 177–223.

Miyashiro, A. 1973. *Metamorphism and Metamorphic Belts*. New York: John Wiley & Sons. 479 pp.

Moore, C. B., and P. P. Sipiera. 1975. *Identification of Meteorites*. Arizona State Univ., Center for Meteorite Studies, Pub. 13., 16 pp.

Moores, E. M. 1973. "Geotectonic significance of ultramafic rocks." *Earth Science Rev.*, v. 9, p. 241–258.

Morgan, B. A. 1975. "Mineralogy and origin of skarns in the Mount Morrison Pendant, Sierra Nevada, California." *Amer. Jour. Science*, v. 275, p. 119–142.

Mount, J. 1985. "Mixed siliciclastic and carbonate sediments: A proposed first-order textural and compositional classification." *Sedimentology*, v. 32, p. 435–442.

Mullineaux, D. R., and D. R. Crandell. 1981. "The eruptive history of Mount St. Helens." In Lipman, P. W., and D. R. Mullineaux (eds). *The 1980 Eruptions of Mount St. Helens, Washington*. U.S. Geol. Survey Professional Paper 1250, p. 3–16.

Palmer, A. R. (Compiler) 1983. "The Decade of North American Geology 1983 Geologic Time Scale." *Geology*, v. 11, p. 503–504.

Papike, J. J. 1988. "Chemistry of the rock-forming silicates: multiple-chain, sheet, and framework structures." *Reviews of Geophysics*, v. 26, p. 407–444.

Park, C., and R. A. MacDiarmid. 1975. *Ore Deposits*. 3rd ed. San Francisco: W. H. Freeman and Co. 530 pp.

Parker, R. L. 1967. "Composition of the Earth's crust." U.S. Geological Survey Prof. Paper. 440D, 14 pp.

Patterson, C. C. 1956. Age of meteorites and the earth. *Geochim. et Cosmochim. Acta*, 10, 230–237.

Powers, M. C. 1953. "A new roundness scale for sedimentary particles." *Jour. Sed. Petrology*, v. 23, p. 117–119.

Press, F. and R. Siever. 1985. *Earth*. 4th ed. San Francisco: W. H. Freeman. 649 pp.

Robbins, M. 1983. *The Collector's Book of Fluorescent Minerals*. New York: Van Nostrand & Reinhold. 289 pp.

Robertson, E. C. 1966. "The Interior of the Earth." *U.S. Geological Survey Circular* 532, 10 pp.

Schweikert, R. A., and D. S. Cowan. 1975. "Early Mesozoic tectonic evolution of the western Sierra Nevada, California." *Geol. Soc. America Bulletin*, v. 86, p. 1329–1336.

Selley, R. C. 1988. *Applied Sedimentology*. New York: Academic Press. 446 pp.

Shay, K. 1975. Mineralogical zoning in a scapolite-bearing skarn body on San Gorgonio Mountain, California. *American Mineralogist*, v. 60, p. 785–797.

Smith, R. L. 1960. "Ash flows." *Geol. Soc. America Bulletin*, 71, p. 795–842.

Smith, R. L., and R. A. Bailey. 1968. *Resurgent Cauldrons*. R. R. Coats, R. L. Hay, and C. A. Anderson (Eds). Boulder: Geological Society of America, Memoir 116.

Streckeisen, A. 1979. "Classification and nomenclature of volcanic rocks, lamprophyres, carbonatites, and melilitic rocks: Recommendations and suggestions of the IUGS Subcommission on the Systematics of Igneous Rocks." *Geology*, v. 7, p. 331–335.

Streckeisen, A. 1976. "To each plutonic rock its proper name." *Earth Science Rev.*, v. 12, p. 1–33.

Sundell, K. A., and R. V. Fisher. 1985. "Very coarse grained fragmental rocks: A proposed size classification." *Geology*, v. 13, p. 692–695.

Swanson, D. A., W. A. Duffield, D. B. Jackson, and D. W. Peterson. 1979. "Chronological Narrative of the 1969–71 Mauna Ulu eruption of Kilauea Volcano, Hawaii." U.S. Geological Survey Prof. Paper 1056, 55 pp.

Tilling, R. I. 1987. "Monitoring Active Volcanoes." *U.S. Geological Survey General Interest Publication*, 13 pp.

Tilling, R. I., R. L. Christiansen, W. A. Duffield, E. T. Endo, R. T. Holcomb, R. Y. Koyanagi, D. W. Peterson, and J. D. Unger. 1987. *The 1972–1974 Mauna Ulu Eruption, Kilauea Volcano: An Example of Quasi-Steady-State Magma Transfer*, R. W. Decker, T. L. Wright, and P. H. Stauffer (Eds.), U.S. Geological Survey Professional Paper 1350, Volcanism in Hawaii, p. 405–470.

Tuttle, M. L. 1991. Geochemical, Biogeochemical, and Sedimentological Studies of The Green River Formation, Wyoming, Utah, and Colorado: *U.S. Geological Survey Bulletin* 1973–A–G.

Warren, J. K. 1989. *Evaporite Sedimentology*. Englewood Cliffs, NJ: Prentice-Hall. 285 pp.

Wentworth, C. K. 1922. "A scale of grade and class terms for clastic sediments." *Jour. Geology*, v. 30, p. 377–392.

Windley, B. F. 1969. "Anorthosites of southern West Greenland." *American Assoc. Petroleum Geol. Mem.* 12, p. 899–915.

Winkler, H. G. F. 1976. *Petrogenesis of Metamorphic Rocks*. 4th ed. New York: Springer-Verlag. 334 pp.

Index

Absolute (chronometric) time scale, 100
Accessory minerals, 41, 126, 138–139
Acicular (needlelike) crystals, 15, 17
Actinolite, 72, 75
Adamantine (diamond-like) luster, 18
Aegerine, 72, 74
Agate, 84, 85
Age of Earth, 94
Agricola, Georgius, 2
Alabaster gypsum, 49
Albite, 83, 85–86
Alkalai feldspars, 83, 86, 123, 125, 126
Alkalic magma, 103
Alkalic rocks, 139–140
Alkali-silica diagram, 122, 123
Allogenic components, 147
Alloys, 5
Almandine, 62, 64
Almandite, 62, 64
Alpine-type ultramafic bodies, 135
Al_2SiO_5 group, 63, 64, 65–66
Amazonstone, 86
Amethyst, 84
Amethyst Vein, 24
Amphibole group, 61, 72–73, 74–75
Amphibolite facies, 175, 179

Amphibolites, 174, 182
Analcime, 83
Andalusite, 64, 65–66
Andesite, 103, 123
Andesite lava, 100
Andradite, 62, 64
Anglesite, 50
Angularity of grain surfaces, 149
Anhydrite, 49, 50
Anions, 10
Anorthite, 83, 85–86
Anorthoclase, 86
Anorthosite massifs, 135
Anorthosites, 135
Anthophylite, 72
Anthracite, 166
Antigorite, 76
Apatite, 55, 56–57
Apparent polar paths, 100
Aquamarine, 69
Aragonite, 13, 45, 47–48
Arenites, 149, 153, 154
Argentite, 30
Argentum (silver), 22–25
Argon, 10
Arkose, 153–154
Armored mudballs, 168, 170
Arsenopyrite, 30
Ash flows and ash-flow tuffs, 118, 120–121
Associations of minerals, 36
Asterism, 40
Asthenosphere, 92, 96

Atolls, coral, 161–162
Atomic number, 8, 196–197
Atomic structure, 7–8
Atomic weight, 8, 196–197
Augen texture, 180, 181
Augite, 71, 72
Authigenic components, 147
Autunite, 55, 58
Axinite, 69
Azurite, 45, 48

Banded-iron formations (BIF), 166, 169
Banded skarn, 178
Barite, 48–49, 50
Basaltic volcanism, 99–100
Batholiths, 127–128
  composite, 128–131
Batopilas District, Mexico, 23–24
Bauxite, 44–45, 142
Bedded cherts, 166
Bedding, 143, 144–146
Belt Series, 35
Bentonite, 79
Beryl, 68, 69–70
BIF (banded-iron formations), 166, 169
Biochemical precipitation, 160, 161
Biotite, 78, 79, 126
Bituminous coal, 166
Bladed crystals, 15, 17
Blasto (prefix and suffix), 180

209

Blastoporphyritic texture, 180
Blueschist facies, 175, 179
Blueschists, 178
Body waves, 91
Boehmite, 38
Borates, 50–53
Borax, 50, 51
Bornite, 30
Botryoidal masses, 43, 44
Bowen's Reaction Series, 134, 193
Bronze, 2
Brucite, 38, 43
Bulbous budding, 122
Bushveld Complex, South Africa, 27, 133–134

Calaverite, 30
Calcalkalic magma, 103
Calcite, 13, 14, 44, 46, 47
Calderas
  collapse, 103, 117–118, 119
  defined, 117
California, Sierra Nevada Batholith, 129–131
Calumet District, Colorado, 177
Caprock, 28
Capulin, Mount (New Mexico), 116
Carbonate platforms, 162–164
Carbonates, 44, 45–48
Carbonatites, 139–140
Carnelian, 84
Carnotite, 55, 58
Cascade Range volcanoes, Washington, 111–115
Cassiterite, 38, 42–43
Cataclastic rocks, 182
Cataclastic texture, 179, 181
Cations, 10
Celestite, 49, 50
Cenozoic era, 188
Cerargyrite, 51
Cerussite, 45, 48
Chabazite, 88
Chalcedony, 84
Chalcocite, 30

Chalcopyrite, 30
Changed rocks. *See* Metamorphism and metamorphic rocks
Chassignites, 95
Chemical and organic sedimentary rocks
  armored mudballs, 168, 170
  classification, 158, 170–171
  concretions, 169–170
  dolostone, 164–165
  evaporites, 159, 165–166
  introduced, 157
  limestone, 14, 159–164, 170, 171
  nodules, 169–170
  other chemical and organic sediments, 166, 167, 169
  seawater, chemical composition of, 157, 159
Chemical bonding, 10–13
Chert (flint), 1, 84
  bedded, 166
Chert lutites, 152
Chill zone, 134
Chimneys, 34
Chlorine, 12
Chlorite, 78, 79
Chromite, 38, 43, 134
Chronometric (absolute) time scale, 100
Chrysoberyl, 43
Chrysocolia, 78
Chrysotile, 76, 77
Cinder cones, 116–117
Cinders, 102, 116
Cinnabar, 30
Citrine, 84
Clay, 156
Clay group, 79
Clay mineralogy of mudrocks, 155, 156
Cleavage, 15–16, 181
Clinoptilolite, 88
Clinopyroxenes (monoclinic pyroxenes), 71, 140
Clinopyroxenite, 140
Coal, 166

Cobaltite, 30
Coesite, 84–85
Colemanite, 51, 53
Collapse calderas, 103, 117–118, 119
Colorado
  Calumet District, 177
  Creede District, 24–25
  Cripple Creek District, 35
  Whitehorn Stock, 177
Color index of rocks, 136
Color of minerals, 15
Columbite, 38
Columnar jointing, 120, 121
Complex (exotic) pegmatites, 131–132
Composite batholiths, 128–131
Composite volcanoes, 100, 110–116
Compositional maturity, 148
Compounds, 10
Conchoidal fracture, 15, 81, 84
Concretions, 169–170
Conglomerates, 26, 149, 150
Contact (thermal) metamorphism, 176–178
Continental evaporites, 165–166
Continental wandering, 100
Copper, 22, 26–27
  porphyry, 3, 4, 26–27
Coquina, 169, 171
Coral atolls, 161–162
Coral reefs, 161–162
Cordierite, 68, 69
Core of Earth, 89–93, 94–95
Corundum, 16, 38, 40–41
Couer d'Alene District, Idaho, 35
Country rocks, 127
Covalent bonding, 12–13
Covellite, 30
Crater Lake, Oregon, 115–116
Creede District, Colorado, 24–25
Crenulations, 181
Cretaceous reef limestone, 160, 161
Cripple Creek District, Colorado, 35

Cristobalite, 80, 82
Crocidolite, 75
Cross-bedding, 145–146
Crust of Earth, 4–5, 89–92, 95–97
Cryolite, 54
Crystal form of minerals, 15, 16, 17
Crystallography, 189–190
Crystals
　acicular (needlelike), 15, 17
　bladed, 15, 17
　fishtail, 49, 52
　hexagonal, 40, 190, 191
　isometric, 41, 190, 191
　monoclinic, 190
　orthorhombic, 190
　silicates, 59–61
　swallowtail, 49, 52
　tabular, 15, 17
　tetragonal, 42, 190
　tetrahedral, 59–61
　triclinic, 190
　trigonal, 47
　twinning of, 67
Cummingtonite, 73
Cuprite, 38
Curie Point, 100
Cyclosilicates, 60, 68, 69–70

Dacite, 103, 123
Deep ocean trenches, 96
Dendritic growths, 42, 43
Deposits
　disseminated, 25
　grade of, 3
　lode, 25
　ore. *See* Ore deposits
　placer, 25
Desulfovibrio desulfuricans bacteria, 28
Detrital sedimentary rocks
　defined, 141
　description and classification of clastic sedimentary rocks, 148–152
　introduced, 141
　mudrocks, 149, 151, 154, 156
　sandstones, 151, 152–155

　sedimentary environments, 142–143
　stories the rocks can tell
　　allogenic and authigenic components, 147
　　diagenesis, 148
　　introduced, 143–144
　　sedimentary structures, 144–146
　weathering and, 141–142
Diagenesis, 148
Diamonds, 22, 28–29, 92–93
Diaspore, 38
Dikes, 127, 128, 129
Diopside, 71, 72
Dispersion of diamonds, 29
Disseminated deposits, 25
Dolomite, 44, 47, 164–165
Dolostone, 44, 47, 164–165
Dunite, 93, 140
Dynamic metamorphism, 179

Earth
　age, 94
　chemical composition of crust, 4
　composition
　　historical perspective, 90–93
　　sources of information about, 89–90
　core, 89–93, 94–95
　crust, 4–5, 89–92, 95–97
　density, average, 89
　magnetic field, 95, 100
　mantle, 89–93, 96–97
　mass, 89
　meteorites and, 90, 93–95
　minerals in crust, 4–5
　seismic waves and, 89–93
　size, 89
Earthquakes, 91
Earthy hematite, 41
Earthy luster, 18
Eclogites, 93, 179
Electrons, 7–8
Electrum, 25
Emeralds, 69

Emery, 41
Emperor Seamount chain, Hawaiian Islands, 101–102
End members, 62
Enstatite, 72
Epicenter of earthquakes, 91
Epidote group, 60, 65, 67–68
Essential minerals, 125, 138
Euhedral diamonds, 29
Eutaxitic texture, 120
Evaporites, 159, 165–166
Exotic (complex) pegmatites, 131–132

Facies, 175, 179–180
Fayalite, 62
Feeder dikes, 128
Feldspar group, 61, 82–83, 85–86
Feldspars, 4, 123, 125, 126
Feldspathoid group, 83, 87
Feldspathoids, rocks free of, 136–139
Fire opal, 84
Fishtail crystals, 49, 52
Fissile mudrocks, 151, 154, 156
Fissures (volcanic), 103
Flank eruptions, 103
Flint (chert), 1, 84
　bedded, 164
Fluorite, 16, 51, 54
Flute casts, 145
Focus of earthquakes, 91
Foliated rocks, 174, 181–183, 184
Fossilferous limestone, 170, 171
Fossils, 143–144
Fosterite, 62
Fracture of minerals, 15
Frasch Process, 28
Fuego Volcano, Guatemala, 110

Galena, 13, 30, 32
Gangue minerals, 36
Garnet, 16
Garnet-chlorite-biotite schist, 182, 183
Garnet group, 59, 62–63, 64

Garnet peridotite, 92
Geologic time scale, 100, 188
Geomagnetic reversals, 100
Geomagnetic time scale, 100
Geotherms, 90
Glass shards, 120
Glaucophane, 73, 75
Gneiss, 182
Goethite, 14, 38, 43, 44
Gold, 22, 25–26
Golden beryl, 69
Grade of deposits, 3
Granite melting curve, 175, 180
Granites and other plutonic rocks
   alpine-type ultramafic bodies, 135
   anorthosites, 135
   classification of plutonic rocks
      introduced, 135–136
      rocks free of feldspathoids, 136–139
      silica unsaturated rocks, 139–140
      ultramafic rocks, 27, 135, 139–140
   composite granitic batholiths, 128–131
   geometry of plutonic bodies, 127–128
   introduced, 14, 127
   pegmatites, 131–133
   stratiform complexes, 27–28, 133–134
Granodiorite, 129
Granulite facies, 175, 180
Granulites, 175–176, 182
Graphic granite, 131, 132
Graphite, 13, 22, 28–29
Greensand, 155
Greenschist facies, 175, 179
Grossularite, 62, 64
Guatemala, Fuego Volcano, 110
Gutenburg Discontinuity, 92
Gypsum, 49, 50, 144

Habit of minerals, 15
Halides, 51, 53–54

Halite (salt), 2, 11, 12, 13, 51, 53, 54
Harding Pegmatite, 71, 74
Hardness of minerals, 17
Harrison, George, 26
Harzburgite, 140
Hawaiian Islands
   Emperor Seamount chain, 101–102
   Kilauea, 103–104, 105, 107–110
   Mauna Kea, 103, 105
   Mauna Loa, 103–104, 105
   Mauna Ulu eruption of Kilauea, 107–110
   shield volcanoes, 103–110
Hawaiian Volcano Observatory, 104
Hedenbergite, 71
Hematite, 38, 41
Herringbone, 23
Hess, Harry, 92, 95
Heulandite, 83, 88
Hexagonal crystals, 40, 190, 191
Hiddenite, 71, 74
Hornblende, 73, 75, 126
Hornfels, 176, 182, 183
Hornfels facies, 175, 179
Hotspot tracers, 101–102
Hydrothermal ore deposits, 29, 32, 33–35
Hydrothermal veins, 2
Hydroxides, 37–39, 43–45

Idaho
   Coeur d'Alene District, 35
   Sunshine Mine, 35
Idocrase (vesuvianite), 60, 65, 68–69
Igneous rocks, 102
   plutonic (intrusive), 127
Ignimbrites, 120
Illite, 155, 156
Ilmenite, 39, 41–42
Inherited texture, 180
Inosilicates. *See under* Silicates
Intraclasts, 163

Intrusive (plutonic) igneous rocks, 127
Ionic bonding, 12
Ions, 9–10
Iron, 22
Iron meteorites, 93, 94
Ironstone concretions, 169–170
Isometric crystals, 41, 190, 191
Isometric minerals, 41
Isotopes, 9

Jadeite, 72, 74
Jasper, 84
Jolly balance, 18

Kamacite, 93
Kaolinite, 155, 156
Kaolinite group, 78, 79
Kernite, 51
Keweenaw Peninsula, Michigan, 26
Kilauea, Hawaiian Islands, 103–104, 105, 107–110
Kimberlite, 28–29, 140
Kunzite, 71
Kyanite, 64, 65–66

Laccoliths, 128
Laminated beds, 144–145
Laminated gypsum, 144
Lamprophyres, 140
Laterites, 142
Latite, 123
Lava, andesite, 100
Lava flows, 121–122
Lepidolite, 78, 79
Leucite, 83, 87
Lherzolite, 140
Lignite, 166
Limestone, 14, 159–164, 170, 171
Lineation, 183, 184
Lithic arenites, 154
Lithosphere, 92, 96
Lizardite, 76–77
Lode deposits, 25
Lodestones, 41
Low velocity zone (LVZ), 92

Luster of minerals, 18
Lutite, 149
LVZ (low velocity zone), 92

Mafic minerals, 136
Magma, 102–103
Magnesite, 44, 47
Magnetic field of Earth, 95, 100
Magnetite, 39, 41, 126
Malachite, 45, 48
Malleable minerals, 21
Manganite, 39
Mantle of Earth, 89–93, 96–97
Mantos, 34
Marble, 182
Marcasite, 30
Marine evaporites, 165
Maturity of a sedimentary process, 148
Mauna Kea, Hawaiian Islands, 103, 105
Mauna Loa, Hawaiian Islands, 103–104, 105
Mauna Ulu eruption of Kilauea, Hawaiian Islands, 107–110
Megabreccia, 118
Mendeleev, Dmitri, 11
Merensky Reef, South Africa, 27, 134
Mesosphere, 96
Mesozoic era, 188
Metallic bonding, 13
Metallic luster, 18
Metallic native elements
  copper, 22, 26–27
    porphyry, 3, 4, 26–27
  gold, 22, 25–26
  introduced, 21
  iron, 22
  platinum, 22, 27–28
  silver (argentum), 22–25
  tabulated, 22
Metamorphism and metamorphic rocks
  classification of metamorphic rocks, 181–185
  introduced, 99, 173
  other types of metamorphism, 178–179
  pressures and temperatures of metamorphism, 175, 179–180
  regional metamorphism, 174–176
  textures and structures, 180–181
  thermal (contact) metamorphism, 176–178
Meteorites, 90, 93–95
Mexico
  Batopilas District, 23–24
  Paricutin Volcano, 117
  Santa Eulalia District, 33–35
  Trans-Mexican Volcanic Belt, 117
Mica group, 77, 79
Michigan, Keweenaw Peninsula, 26
Micrite, 170
Microcline, 82, 86
Mid-ocean ridges, 95, 99–100
Migmatite, 174
Millerite, 31
Minerals. *See also* Ore deposits
  accessory, 41, 126, 138–139
  associations, 36
  chemical classification, 19
  cleavage, 15–16, 181
  color, 15
  crystal form, 15, 16, 17
  defined, 1, 13
  in Earth's crust, 4–5
  essential, 125, 138
  fracture of, 15
  gangue, 36
  habit, 15
  hardness, 17
  introduced, 1, 13–14
  isometric, 41
  luster of, 18
  mafic, 136
  malleable, 21
  mode (volume percent composition of minerals), 136–138
  number of, 10
  physical properties, 14–19
  polymorphs, 13
  precipitation of, 159
  pseudomorphs, 13–14
  rock-forming. *See* Silicates
  secondary, 139
  specific gravity, 18
  stoichiometric, 13
  streak of, 18
  use by early civilizations, 2
Mode (volume percent composition of minerals), 136–138
Mohoricivic Discontinuity, 92, 130
Moh's Hardness Scale, 17
Molybdates, 55, 56
Molybdenite, 31
Monazite, 55, 57
Monoclinic carbonates, 45, 48
Monoclinic crystals, 190
Monoclinic pyroxenes (clinopyroxenes), 71, 140
Monomineralic rocks, 14, 102
Montana, Stillwater Complex, 28, 133, 134
Montmorillonite, 155, 156
Montmorillonite clay, 79
Morganite, 69
Mount Capulin, New Mexico, 116
Mount Saint Helens, Washington, 111–115
Mudcracks, 146, 147
Mudrocks, 149, 151, 154, 156
Muscovite, 77, 78, 79
Mylonite, 179, 182

Nakhlites, 95
Native elements
  metallic. *See* Metallic native elements
  nonmetallic, 22, 28–29
Natrolite, 88
Nepheline, 83, 87, 125

Nesosilicates. *See under* Silicates
Neutrons, 7–8
New Mexico
    Mount Capulin, 116
    Permian Basin, 165
    Raton-Clayton Volcanic Field, 104
Niccolite, 31
Nodules, 169–170
Nonfissile mudrocks, 154, 156
Nonfoliated rocks, 182, 183–185
Nonmetallic luster, 18
Nonmetallic native elements, 22, 28–29
Novaculite, 166
Nuclear force, 8
Nucleus of atoms, 7–8

Obsidian, 1, 123, 124, 125
Octahedrites, 93, 94
Olivine, 13, 90, 92–93, 140
Olivine group, 59, 62, 64
Olivine-pyroxene (peridotite-like) composition, 90
Oncolites, 163
Oölids, 162–163
Opal, 80, 82, 84
Ore deposits
    chimneys, 34
    defined, 3
    hydrothermal, 29, 32, 33–35
    mantos, 34
    quest for, 2–4
Oregon, Crater Lake, 115–116
Organic sedimentary rocks. *See* Chemical and organic sedimentary rocks
Orpiment, 31
Orthoclase, 82, 86
Orthopyroxenes (orthorhombic pyroxenes), 71, 140
Orthopyroxenite, 140
Orthorhombic carbonates, 45, 47–48
Orthorhombic crystals, 190
Orthorhombic pyroxenes (orthopyroxenes), 71, 140

Oxidation (rusting), 5
Oxides
    cassiterite, 38, 42–43
    corundum, 16, 38, 40–41
    defined, 37
    hematite, 38, 41
    ilmenite, 39, 41–42
    magnetite, 39, 41, 126
    other, 43
    rutile, 39, 41–42
    tabulated, 38–39

Pahoehoe, 122
Paleozoic era, 188
Paricutin Volcano, Mexico, 117
Partial melting zone, 92, 96–97
Pearly (satiny/silky) luster, 18
Pectolite, 73, 76
Pegmatites, 131–133
Pellets, 163
Pentlandite, 31
Peralkalic magma, 103
Peridotite, 92, 140
Peridotite-like (olivine-pyroxene) composition, 90
Periodic Table, 8, 11–12, 196–197
Perlite, 123
Permian Basin, New Mexico and Texas, 165
Perovskite, 90
Phacoliths, 128
Phase diagrams, 62, 63, 80
Phenocryst minerals, 123, 124, 125
Phlogopite, 78, 79
Phosphates, 55, 56–58
Phosphatic units, 166, 167
Phosphorite, 166, 167
Phyllarenites, 152
Phyllite, 181, 182
Phyllosilicates. *See under* Silicates
Pigeonite, 71
Pillow lavas, 122
Pisolites, 162, 163
Placer deposits, 25
Placer sand, 155

Plagioclase feldspars, 83, 85–86, 123, 125, 126
Plastics, 5
Plates, 95
Plate tectonics, 95–96, 99–100
Platinum, 22, 27–28
Plinian eruption column, 114
Plugs, 127
Plutonic (intrusive) igneous rocks, 127
Plutonic rocks. *See* Granites and other plutonic rocks
Plutons, 127, 128
Poise (unit of viscosity), 102
Polar wandering curves, 100
Polymorphs, 13
Porphyroblastic texture, 180
Porphyroblasts, 180
Porphyry copper, 3, 4, 26–27
Precambrian era, 188
Precambrian shields, 176
Precipitation
    biochemical, 160, 161
    of minerals, 159
Primary (P) waves, 91–93
Protolith, 173
Protons, 7–8
Pseudomorphs, 13–14
Psilomelane, 39, 44
Pumice, 123
P (primary) waves, 91–93
Pyralspite group, 62
Pyrite, 14, 31, 33
Pyroclastic sand, 152
Pyrolusite, 39, 43
Pyrope, 62, 64
Pyrophylite, 78
Pyroxene, 90, 92–93
Pyroxene group, 61, 71, 72, 74
Pyroxenite, 92, 140
Pyroxenoid group, 73, 75–76
Pyrrhotite, 31

Quartz, 4, 16, 123, 125
Quartz arenite, 153
Quartz (silica) group, 61, 80–82, 84–85

Index

Quartzite, 182
Quartz monzonite, 129

Radioactive isotopes, 9, 100
Radiometric dating, 9, 100
Raton-Clayton Volcanic Field, New Mexico, 104
Realgar, 31
Reefs, 26
   coral, 161–162
Refractive index of diamonds, 29
Refractory materials, 62
Regional metamorphism, 174–176
Relative hardness, 17
Relict textures, 173
Remanant magnetism, 100
Resinous (greasy) luster, 18
Rhodocrosite, 44, 47, 76
Rhodonite, 73, 76
Rhombohedral carbonates, 44, 45–47
Rhombohedral cleavage, 47
Rhyodacite, 123
Rhyolites, 102, 103, 118, 123
Rhyolitic composition, 124, 125
Riebeckite, 73, 75
Ripple marks, 146, 147
Rock-forming minerals. *See* Silicates
Rocks
   alkalic, 139–140
   caprock, 28
   cataclastic, 182
   changed. *See* Metamorphism and metamorphic rocks
   color index, 136
   country, 127
   defined, 14, 102
   foliated, 174, 181–183, 184
   igneous, 102, 127
   metamorphic. *See* Metamorphism and metamorphic rocks
   monomineralic, 14, 102
   non-foliated, 182, 183–185
   plutonic. *See* Granites and other plutonic rocks
   radiometric dating of, 9, 100
   sedimentary, 99. *See also* Chemical and organic sedimentary rocks; Detrital sedimentary rocks
   silica unsaturated, 139–140
   ultramafic, 27, 135, 139–140
   volcanic. *See* Volcanism and volcanic rocks
Rose quartz, 84
Roundness of grain surfaces, 149
Rudite, 149
Rusting (oxidation), 5
Rutilated quartz, 84
Rutile, 39, 41–42

Saint Helens, Mount (Washington), 111–115
Salt (halite), 2, 11, 12, 13, 51, 53, 54
Salt domes, 28
Sandstones, 151, 152–155
Sanidine, 82, 86
Santa Eulalia District, Mexico, 33–35
Saturated solution, 159
Scheelite, 54, 55
Schist, 182
Schistocity, 181
Scolecite, 83, 87, 88
Scoria, 123, 125
Seamounts, 101
Seawater, 157, 159
Secondary minerals, 139
Secondary (S) waves, 91–93
Sedimentary rocks, 99. *See also* Chemical and organic sedimentary rocks; Detrital sedimentary rocks
Seismographs, 107
Seismology, 89–93
Selenite gypsum, 49, 52
Septarian nodules, 168, 169
Serpentine group, 76–77, 78
Sets (erosionally bounded packages), 146
Shale, 151, 154, 156, 181
Shear zones, 179
Shergottites, 95
Shield volcanoes, 103–110
Siderite, 44, 47
Sierra Nevada Batholith, California, 129–131
Silica (quartz) group, 61, 80–82, 84–85
Silicate magma, 103
Silicates
   crystal structure, 59–61
   cyclosilicates, 60, 68, 69–70
   in Earth's crust, 4
   inosilicates, 60, 61, 70, 72–73
      amphibole group, 61, 72–73, 74–75
      pyroxene group, 61, 71, 72, 74
      pyroxenoid group, 73, 75–76
   introduced, 59
   nesosilicates, 59, 60, 64–65
      Al_2SiO_5 group, 63, 64, 65–66
      garnet group, 59, 62–63, 64
      olivine group, 59, 62, 64
   phyllosilicates, 60, 61, 76, 78
      clay group, 79
      mica group, 77, 79
      serpentine group, 76–77, 78
   sorosilicates, 60, 65
      epidote group, 60, 65, 67–68
      vesuvianite (idocrase), 60, 65, 68–69
   tectosilicates, 60, 61, 80, 82–83
      feldspar group, 61, 82–83, 85–86
      feldspathoid group, 83, 87
      quartz (silica) group, 61, 80–82, 84–85
      zeolite group, 83, 87–88
Silica unsaturated rocks, 139–140
Sillimanite, 64, 65–66
Sills, 128
Silver (argentum), 22–25
Silver Act, 24

Simple pegmatites, 131
Simthsonite, 44, 47
Skarns, 177–178, 182
Skylights (holes in lava tube roofs), 108
Slate, 181, 182
Smectite group, 79
Smoky quartz, 84
Sole marks, 145
Solid solutions, 62
Solidus, 62
Sorosilicates. See under Silicates
Sorting of grain size, 148–149
South Africa
    Bushveld Complex, 27, 133–134
    Merensky Reef, 27, 134
    Witwatersrand Basin, 26
Sparite, 170–171
Specific gravity of minerals, 18
Specular hematite, 41
Spessartite, 62, 64
Sphalerite, 31
Sphene (titanite), 65, 67
Spinel, 39, 90
Spodumene, 71, 72
Stable isotopes, 9
Stainless steels, 5
Stain spar gypsum, 49
Staurolite, 65, 67
Steels, 5
Stibnite, 31
Stilbite, 83, 88
Stillwater Complex, Montana, 28, 133, 134
Stishovite, 84–85
Stocks, 127
Stoichiometric minerals, 13
Stony-iron meteorites, 93
Stony meteorites, 93–94
Stratiform complexes, 27–28, 133–134
Streak of minerals, 18
Stream tin, 42
Strontianite, 45, 48
Subduction zones, 96, 130, 131
Sulfates, 48–49, 50

Sulfides
    introduced, 29, 32
    physical properties, 32–33
    tabulated, 30–31
Sulfosalts
    introduced, 29, 32
    physical properties, 32–33
    tabulated, 30–31
Sulfur, 22, 28
Sunshine Mine, Idaho, 35
Surface waves, 91
Swallowtail crystals, 49, 52
S (secondary) waves, 91–93
Swelling clays, 79
Sylvanite, 31
Sylvite, 51, 53–54

Tabular crystals, 15, 17
Taenite, 93
Tectosilicates. See under Silicates
Tellurides, 33
Terrigeneous sands, 152
Tetragonal crystals, 42, 190
Tetrahedral crystals, 59–61
Texas, Permian Basin, 165
Textural maturity, 148
Textures
    augen, 180, 181
    blastoporphyritic, 180
    cataclastic, 179, 181
    eutaxitic, 120
    inherited, 180
    lava flows, 121–122
    metamorphic rocks, 180–181
    porphyroblastic, 180
    relict, 173
Thermal (contact) metamorphism, 176–178
Tigers eye, 75
Tiltmeters, 105, 106
Time scales
    absolute (chronometric), 100
    geologic, 100, 188
    geomagnetic, 100
Tin, 42–43
Titanite (sphene), 65, 67
Topaz, 65, 66

Tourmaline, 68, 69, 70
Trachyte, 123
Trans-Mexican Volcanic Belt, Mexico, 117
Tremolite, 73, 75
Triclinic crystals, 190
Tridymite, 80, 82
Trigonal crystals, 47
Tuffs, 120, 123
Tungstates, 54–56
Turbidity current, 145
Turquoise, 55, 57
Twining of crystals, 67

Ugrandite group, 62
Ulexite, 51, 52
Ultramafic rocks, 27, 135, 139–140
Ultramylonite, 179
Union of South Africa. See South Africa
Uraninite, 39, 43
Uvarovite, 62, 64

Van der Waals bonding, 13
Very low grade metamorphism, 178
Vesuvianite (idocrase), 60, 65, 68–69
Viscosity, 102
Vitreous (glassy) luster, 18
Vitrophyre, 123, 125
Volcanic bombs, 117
Volcanism and volcanic rocks
    ash flows and ash-flow tuffs, 118, 120–121
    classification of volcanic rocks, 122, 123–126
    introduced, 99
    lava flows, 121–122
    magma, 102–103
    monitoring and predicting eruptions, 104–107
    scientific implications, 99–102
    types of volcanoes and vent areas
        cinder cones, 116–117

Index

collapse calderas, 103, 117–118, 119
composite volcanoes, 100, 110–116
  introduced, 103
  shield volcanoes, 103–110
volcanic rocks defined, 99

Wackes, 151
Walker, George, 26
Washington
  Cascade Range volcanoes, 111–115

Mount Saint Helens, 111–115
Wavellite, 55, 57, 58
Weathering, 141–142
Websterite, 140
Wehrlite, 140
Welded tuffs, 120, 123
Wentworth (1922) size scale, 149, 150
Whitehorn Stock, Colorado, 177
Widmanstatten Structure, 93, 94
Witherite, 45, 48

Witwatersrand Basin, South Africa, 26
Wolframite, 54–56
Wollastonite, 73, 76
Womack, Bob, 35
Wulfenite, 55, 56

Xenoliths, 92

Zeolite group, 83, 87–88
Zincite, 39
Zircon, 63–64, 65

ISBN 0-02-320452-4

9 780023 204524

90000>